Lecture Notes in Mathematics

Edited by A. Dold and B. Ec

1238

Michael Holz
Klaus-Peter Podewski
Karsten Steffens

Injective Choice Functions

Springer-Verlag
Berlin Heidelberg New York London Paris Tokyo

Authors

Michael Holz
Klaus-Peter Podewski
Karsten Steffens
Institute of Mathematics, University of Hannover
Welfengarten 1, 3000 Hannover, Federal Republic of Germany

Mathematics Subject Classification (1980): 04-02, 05-02, 03E05, 04A20, 05A05, 05C70

ISBN 3-540-17221-1 Springer-Verlag Berlin Heidelberg New York
ISBN 0-387-17221-1 Springer-Verlag New York Berlin Heidelberg

Printed in Germany

Printing and binding: Druckhaus Beltz, Hemsbach/Bergstr.
2146/3140-543210

Preface

A marriage of a family F of sets is an injective choice function for F. The marriage problem consists in establishing necessary and sufficient criteria which decide if a family has an injective choice function. First P. Hall formulated his well-known criterion for finite families in 1935. This criterion was generalized by M. Hall to infinite families which have finite members only. A detailed discussion of the results up to 1970 and many applications can be found in Mirsky's book [Mi]. In the seventies the research on the marriage problem took a rapid development. Several necessary and sufficient conditions for countable families were found ; on the one hand transfinite versions of Hall's Theorem, as for example in [N2], on the other hand extensions of the Compactness Theorem as in [HPS 1]. In Chapter III we are going to present these criteria and show that they are all equivalent.

But only three years ago, R. Aharoni, C.St.J.A. Nash-Williams, and S. Shelah published the first necessary and sufficient criterion for arbitrary families. Its form follows the one of P. Hall's Theorem: a family has a marriage if and only if it does not contain one of a set of "forbidden" substructures. Similar criteria can be found in the second chapter of this book.

The Aharoni - Nash-Williams - Shelah - criterion, which we obtain as a consequence of a criterion of K.P. Podewski in this book, has been successfully applied by Aharoni to solve some famous problems in graph theory. His main result is the proof of a strong form of König's Duality Theorem, suggested by P. Erdös. As a consequence he could prove a strong version of Menger's Theorem for graphs which contain no infinite path. One aim of this book is a self-contained representation of these intricate theorems. For this reason we have inserted a separate chapter on set theory for those readers who are not so familiar with transfinite methods. We suggest reading the introduction after the study of Chapter I.

Hannover, March 1986

CONTENTS

Introduction

A family $F = (F(i) : i \in I)$ is a function with domain I which assigns to each $i \in I$ a set $F(i)$. If J is a subset of I, we define $F(J) = \cup\{F(i) : i \in J\}$. A choice function for F is a function $f : I \to F(I)$ such that $f(i) \in F(i)$ for any $i \in I$. We call every injective choice function for F a marriage of F. The aim of this book is to characterize the families which possess a marriage. In the introduction we want to give some examples of families which do not have a marriage and using these examples to deduce conditions for the existence of a marriage whose sufficiency we shall prove in later chapters. Of course, we have to use some set theoretic notions and theorems - they can be found in Section I.1. [1)]

A. First let us consider families $F = (F(i) : i \in I)$ such that $F(I)$ is finite. It is easy to specify a family of this kind which has no marriage:

Example I: Let $I = \{0,1,2\}$ and $F(0) = F(1) = \{0\}$, $F(2) = \{0,1,2\}$. F cannot possess a marriage, since $|F(\{0,1\})| < |\{0,1\}|$.

P. Hall has shown that a family $(F(i) : i \in I)$ such that $F(I)$ is finite has a marriage if and only if the following condition is satisfied:

(i) $|F(J)| \geq |J|$ for any $J \subseteq I$.

It is obvious that condition (i) is necessary for the existence of a marriage for arbitrary families. In Section III.1 we investigate families F such that $|F(i)| < \aleph_0$ for any $i \in I$, so-called Hall families, and we prove that F has a marriage if condition (i) is fulfilled for each finite subset J of I.

---

1) Chapters are denoted by Roman numerals. Section I.1 is Section 1 of Chapter I. The eighth numbered enunciation of I.1 is I.1.8.
If we refer to it in the same chapter, we write 1.8.

B. Next we consider the case that $|F(I)| \leq \aleph_0$. As mentioned above, (i) has to be fulfilled, especially I must be countable. The following family has property (i) but possesses no marriage:

Example II: Let $I = \omega + 1$, $F(n) = \{n\}$ for each $n \in \omega$, and $F(\omega) = \omega$.

So we have to look for further conditions. We call a subset K of I critical in F if $F \upharpoonright K = (F(i) : i \subset K)$ has a marriage and if the range of any marriage f of $F \upharpoonright K$ is $F(K)$ (we write $\text{rng}\, f = F(K)$). If the family F possesses a marriage, then obviously the following condition must be satisfied:

(i') For each $K \subseteq I$ critical in F and each $i \in I \setminus K$, $F(i) \not\subseteq F(K)$.

The family of Example II does not have this property: Take $K = \omega$ and $i = \omega$, then K is critical in F, $i \in I \setminus K$, and $F(i) \subseteq F(K) = \omega$.

In Section III.3 we show that countable families, those are families such that $|I| \leq \aleph_0$, possess a marriage if they have property (i'). This implies that a family F such that F(I) is countable has a marriage if and only if (i) and (i') are satisfied by F.

In the literature you can find a couple of necessary and sufficient conditions for the existence of a marriage for a countable family. We deal with these criteria in Chapter III. There we show that a family F has a property required by any of the criteria if and only if it has property (i'). Further we give several characterizations of critical sets in that chapter which do not use the notion of a marriage.

C. We now discuss the case that $|F(I)| \leq \aleph_1$ and $|F(i)| < \aleph_1$ for any $i \in I$. Clearly conditions (i) and (i') must be fulfilled in this case, too if F possesses a marriage. But they are not sufficient for the existence of a marriage. The most popular example is the following:

Example III: Let $L(\omega_1) = \{\alpha < \omega_1 : \alpha \text{ is a limit ordinal}\}$, let $I = L(\omega_1)$ and $F(\alpha) = \alpha$ for any $\alpha \in I$. Since $|F(i)| \geq \aleph_0$ for any $i \in I$, it follows immediately that (i) holds. (i') is less obvious. But the following lemma shows that $K = \emptyset$ is the only critical subset of I, and so (i') is satisfied. At the same time this lemma is the first step to the characterization of critical sets in Section III.6. [1)]

---

1) Each lemma in this introduction will be reproved in later sections.

<u>Lemma 1</u>. If $K \subseteq I$ is critical in $F = (F(i) : i \in I)$ and $K \neq \emptyset$, then there exists an $i \in K$ such that $|F(i)| < \aleph_0$.

<u>Proof</u>. Assume for contradiction that $|F(i)| \geq \aleph_0$ for any $i \in K$ and let f be a marriage of $F \upharpoonright K$. For each $n \in \omega$ we choose $i_n \in K$ as follows: Let $i_0 \in K$. If i_n is defined for any $k \leq n$, there exists an $a \in F(i_n) \setminus \{f(i_k) : k \leq n\}$, since $|F(i_n)| \geq \aleph_0$. We have $f[K] = F(K)$, since $f \upharpoonright K$ is critical, so there exists an $i_{n+1} \in K$ with $f(i_{n+1}) = a$. Observe that $i_{n+1} \notin \{i_k : k \leq n\}$, since f is injective. Consequently the function $h = (f \setminus \{(i_n, f(i_n)) : n \in \omega\}) \cup \{(i_n, f(i_{n+1})) : n \in \omega\}$ is a marriage of $F \upharpoonright K$ such that $f(i_0) \in F(K) \setminus \operatorname{rng} h$, which contradicts the fact that K is critical in F.

Coming back to our example we point out that F has no marriage, though conditions (i) and (i') hold for F. To see this you need some set theory, especially Fodor's Theorem (Theorem I.6.8). Every choice function for F is regressive on the stationary set $L(\omega_1)$, and such a function cannot be injective by Fodor's Theorem.

Example III leads us to the formulation of a new condition. For that reason let us call a sequence $(A_\alpha : \alpha < \omega_1)$ a special test if $|A_\alpha| < \aleph_1$ and $A_\alpha = \bigcup\{A_{\beta+1} : \beta < \alpha\}$ for any $\alpha < \omega_1$ (the second property implies that the sequence is increasing and continuous). Now we formulate

(ii) For each special test $(A_\alpha : \alpha < \omega_1)$ the set
$S = \{\alpha : \exists i \in I\ (F(i) \subseteq A_\alpha \wedge \forall \beta < \alpha\ (F(i) \not\subseteq A_\beta))\}$
is not stationary in ω_1.[1)]

Example III violates this condition: Take $A_\alpha = \alpha$ for any $\alpha < \omega_1$. Then $S = L(\omega_1)$, and this set is stationary in ω_1.

For the first time the necessity of our condition is not obvious.

<u>Lemma 2</u>. If a family F has a marriage, then it fulfils property (ii).

<u>Proof</u>. Let f be a marriage of F and suppose, to get a contradiction, that there exists a special test $(A_\alpha : \alpha < \omega_1)$ such that the corresponding set S is stationary in ω_1. Without loss of generality we can assume that $A_\alpha \subseteq \omega_1$ for any $\alpha < \omega_1$. $S \cap L(\omega_1)$ is

---

1) Remember that you can find the set theoretic definitions in Section I.1.

stationary, too. To get a regressive function h, choose, for each $\alpha \in S \cap L(\omega_1)$, an i_α such that $F(i_\alpha) \subseteq A_\alpha$ and $F(i_\alpha) \not\subseteq A_\beta$ for any $\beta < \alpha$. Since α is a limit ordinal, we have $\bigcup\{A_\beta : \beta < \alpha\} = A_\alpha$, and so there exists some $h(\alpha) < \alpha$ with $f(i_\alpha) \in A_{h(\alpha)}$. By Fodor's Theorem, there exist a $\gamma \in \omega_1$ and a stationary subset S_0 of S such that $h(\alpha) = \gamma$ for each $\alpha \in S_0$. Then $f(i_\alpha) \in A_\gamma$ for all $\alpha \in S_0$. Now clearly $i_\alpha \neq i_\beta$ for $\alpha \neq \beta$, so $|\{i_\alpha : \alpha \in S_0\}| = |S_0| = \aleph_1 = |\{f(i_\alpha) : \alpha \in S_0\}| \leq |A_\gamma|$, which contradicts the fact that $|A_\gamma| < \aleph_1$. This completes the proof of the lemma.

But even the conjunction of conditions (i), (i'), and (ii) still is not sufficient for the existence of a marriage of a family $F = (F(i) : i \in I)$ with $|F(I)| \leq \aleph_1$. To see this, regard

Example IV: Let $I = L(\omega_1) \cup \{\alpha+1 : \alpha \in L(\omega_1)\}$ and define $F(\alpha) = F(\alpha+1) = \alpha+1$ for each $\alpha \in L(\omega_1)$. Since $|F(i)| = \aleph_0$ for each $i \in I$, (i) and (i') are satisfied as it has been shown above. To prove that F has property (ii), let $(A_\alpha : \alpha < \omega_1)$ be a special test, and let S be the corresponding set. We have to show that $S \cap C = \emptyset$ for some closed unbounded subset C of $\aleph_1$. Take $J = \{i \in I : \exists \alpha < \omega_1 (F(i) \subseteq A_\alpha)\}$. If J is countable, then there exists an $\alpha_0 < \omega_1$ such that $F(i) \subseteq A_{\alpha_0}$ for any $i \in J$, since the sequence $(A_\alpha : \alpha < \omega_1)$ is increasing. $C = \{\alpha \in \omega_1 : \alpha_0 < \alpha\}$ has the desired properties. Now let $|J| = \aleph_1$. By the definition of F and J, we can find, for each $\beta < \omega_1$, an $\alpha < \omega_1$ such that $\alpha > \beta$ and $\beta \subseteq A_\alpha$. On the other hand each A_α is contained in some $\beta > \alpha$, since A_α is countable. So we can define, for any $\gamma < \omega_1$, an increasing sequence $(\alpha_n : n \in \omega)$ such that $\alpha_0 > \gamma$ and $\alpha_0 \subseteq A_{\alpha_1} \subseteq \alpha_2 \subseteq A_{\alpha_3} \subseteq \dots$.

If $\alpha = \sup\{\alpha_{2i} : i \in \omega\} = \sup\{\alpha_{2i+1} : i \in \omega\}$, then $\alpha = \bigcup\{\alpha_{2i} : i \in \omega\} = \bigcup\{A_{\alpha_{2i+1}} : i \in \omega\} = A_\alpha$, since the sequence $(A_\alpha : \alpha < \omega_1)$ is continuous. So the set $C = \{\alpha \in L(\omega_1) : A_\alpha = \alpha\}$ is unbounded in ω_1 and, again by continuity, it is closed. Since every F(i) is a successor ordinal, we have $C \cap S = \emptyset$, and property (ii) is established for F.

Nevertheless F has no marriage because, for each choice function f for F, the function $h : L(\omega_1) \to \aleph_1$, defined by $h(\alpha) = \min\{f(\alpha), f(\alpha+1)\}$ for each $\alpha < \omega_1$, is regressive on the stationary set $L(\omega_1)$. Consequently h is not injective, and it easily follows that f is not injective.

This example leads to a further condition. Define $F^A = (F(i) : F(i) \subseteq A)$ and $F_A = (F(i) \setminus A : F(i) \not\subseteq A)$.

(iii) If $(A_\alpha : \alpha<\omega_1)$ is a special test, then the set

$$T = \{\alpha<\omega_1 : \exists\beta>\alpha\,(F_{A_\alpha}^{A_\beta} \text{ has no marriage})\} \text{ is not stationary.}$$

The family in Example IV does not meet condition (iii): Take $A_\alpha = \alpha$ for each $\alpha<\omega_1$. Then $F_{A_\alpha}^{A_{\alpha+1}} = \{(\alpha,\{\alpha\}),(\alpha+1,\{\alpha\})\}$ has no marriage for any element α of the stationary set $L(\omega_1)$. So the corresponding set $T \supseteq L(\omega_1)$ is stationary. Again we need Fodor's Theorem to show that (iii) is necessary for the existence of a marriage.

<u>Lemma 3</u>. If a family F has a marriage, then it satisfies property (iii).

<u>Proof</u>. Let f be a marriage of F and assume, to get a contradiction, that there exists a special test $(A_\alpha : \alpha < \omega_1)$ such that the corresponding set T in property (iii) is stationary. To get a regressive function h choose, for each element α of the stationary set $T\cap L(\omega_1)$, an ordinal $\beta>\alpha$ such that $F_{A_\alpha}^{A_\beta}$ has no marriage. Now $f \upharpoonright \{i\in I : F(i)\subseteq A_\beta\}$ is a marriage of F^{A_β}, therefore there must exist an i_α with $F(i_\alpha)\not\subseteq A_\alpha$ such that $F(i_\alpha)\subseteq A_\beta$ and $f(i_\alpha)\in A_\alpha$. Since α is a limit ordinal and consequently $A_\alpha = \bigcup\{A_\beta : \beta<\alpha\}$, there exists an ordinal $h(\alpha)<\alpha$ with $f(i_\alpha)\in A_{h(\alpha)}$. By Fodor's Theorem, there exist an ordinal $\gamma<\omega_1$ and a stationary subset T_0 of $T\cap L(\omega_1)$ such that $h(\alpha) = \gamma$ for each $\alpha\in T_0$. This implies $f(i_\alpha)\in A_\gamma$ for each $\alpha\in T_0$, and, since f is injective and $|A_\gamma| < \aleph_1$, we get a contradiction if we can show that $|\{i_\alpha : \alpha\in T_0\}| = \aleph_1$. If $|\{i_\alpha : \alpha\in T_0\}| \leq \aleph_0$, then, by the definition of the i_α's, there exists an ordinal $\beta<\omega_1$ such that $F(i_\alpha)\subseteq A_\beta$ for each $\alpha\in T_0$. On the other hand T_0 is stationary and consequently unbounded. So we can find a $\sigma\in T_0$ with $\sigma>\beta$. Now $A_\beta\subseteq A_\sigma$ and $F(i_\sigma)\subseteq A_\sigma$ which contradicts the choice of i_σ. Thus $|\{i_\alpha : \alpha\in T_0\}| = \aleph_1$, and our assumption that T is stationary has led to a contradiction. This completes the proof of Lemma 3.

Shelah has proved in [S1] that a family $F = (F(i) : i\in I)$ with the properties $|F(I)| \leq \aleph_1$ and $|F(i)| < \aleph_1$ for each $i\in I$ has a marriage if it satisfies the conditions (i), (ii), (iii), and additionally

(iv) For any countable set A, the family F^A has a marriage.

We have illustrated in case B that we can replace (iv) by (i'). Since, for any special test $(A_\alpha : \alpha < \omega_1)$, the sets $A_\beta \setminus A_\alpha$ are countable, we can replace for the same reason in (iii) the statement "$F_{A_\alpha}^{A_\beta}$ has no marriage" by "$F_{A_\alpha}^{A_\beta}$ does not fulfil (i')". So we obtain for case C a necessary and sufficient condition for the existence of a marriage.

D. The next generalization consists in dropping the requirement "$|F(i)| < \aleph_1$ for each $i \in I$". So we discuss families $F = (F(i) : i \in I)$ which have the property $|F(I)| \leq \aleph_1$. Now the conjunction of (i), (i'), (ii), and (iii) is not sufficient for the existence of a marriage.
<u>Example V</u>: Let $I = \omega_1 + \omega_1$, and define $F(\alpha) = \omega \cup \{\alpha\}$ for $\alpha \in \omega_1$ and $F(\alpha) = \omega_1$ for $\alpha \in I \setminus \omega_1$. (i) is satisfied for F. If $(A_\beta : \beta < \omega_1)$ is a special test, then $\{i \in I : \exists \beta < \omega_1 (F(i) \subseteq A_\beta)\} \subseteq \omega_1$. Since $F \upharpoonright \omega_1$ has a marriage, property (ii) and property (iii) follow respectively by Lemma 2 and Lemma 3 for $F \upharpoonright \omega_1$ and so for F. Further $|\{i \in I : F(i) \subseteq A\}| \leq \aleph_0$ for each $i \in I$, so (iv) is fulfilled. If f is a marriage of F, the set $J = f[I \setminus \omega_1]$ has cardinality $\aleph_1$. But if $\alpha \in J$, then $f(\alpha) \in \omega$, since $\alpha = f(\beta)$ for some $\beta > \omega_1$. So $f[J] \subseteq \omega$ and $|f[J]| = \aleph_1$. This contradiction shows that F possesses no marriage. We observe the following phenomenon: $K = \omega_1 \setminus \omega$ is critical in F_ω and $|\{i \in I \setminus K : F(i) \subseteq F(K) \cup \omega\}| > |\omega|$.

This leads to a new requirement whose necessity for the existence of a marriage cannot be seen so easily and which is one of the essential properties of critical sets which are discussed in Section II.2.

(1) For each set A and each K critical in F_A,
$|\{i \in I \setminus K : F(i) \subseteq F(K) \cup A\}| \leq |A|$.

If a family satisfies (1), it satisfies (i) and (i'). Roughly speaking, condition (1) is condition (i) "modulo critical sets". An important role will be played by <u>the essential size</u> of a family F, denoted by $\|F\|$:
$\|F\| = \min\{|A| : \exists K (K \text{ is critical in } F_A \text{ and } F(I) = F(K) \cup A)\}$.

Again we can say that $|F(I)|$ is $\|F\|$ modulo critical sets. If $\|F\| < \aleph_0$, one could say that F is a finite Hall family modulo critical sets. If for each $i \in I$ there exists a $J \subseteq I$ such that $i \in J$ and $\|F \upharpoonright J\| < \aleph_0$, we call the family F a stable family. Stable families behave like Hall families modulo critical sets and are discussed in Section III.7.

We show in Section II.4 that a family F with $\|F\| \leq \aleph_0$ possesses a marriage if and only if (1) is satisfied. If $|F(I)| \geq \aleph_1$, the conjunction of (1), (ii), and (iii) does not guarantee the existence of a marriage. To see this, consider the following examples.

Example VI: Let $I = \omega_1 \cup \{\omega_1 + \alpha : \alpha \in L(\omega_1)\}$ and define $F(\alpha) = \{\alpha\}$ if $\alpha < \omega_1$, $F(\alpha) = \alpha$ if $\alpha \in I \setminus \omega_1$.

Example VII: Let $I = \omega_1 \cup \{\omega_1 + \alpha : \alpha \in L(\omega_1)\} \cup \{\omega_1 + \alpha + 1 : \alpha \in L(\omega_1)\}$ and define $F(\alpha) = \{\alpha\}$ if $\alpha < \omega_1$, $F(\omega_1 + \alpha) = \omega_1 + \alpha + 1 = F(\omega_1 + \alpha + 1)$ otherwise.

For both examples, conditions (1), (ii), and (iii) are satisfied, as the reader may verify for himself after the study of II.2, using II.2.5. Analogously to the procedure for Examples III and IV one can show that the families do not possess a marriage.

In the same way as we have generalized condition (i) "modulo critical sets" to condition (1), we will proceed with (ii) and (iii). Let μ be an uncountable regular cardinal. A continuous chain of sets $(X_\alpha : \alpha < \mu)$ is called a μ-test of F if, for each $\alpha < \mu$, there exist a set A_α and a set K_α critical in $F \setminus A_\alpha$ such that $|A_\alpha| < \mu$ and $X_\alpha = A_\alpha \cup F(K_\alpha)$. Every $\aleph_1$-test is a special test - choose $K_\alpha = \emptyset$ for each $\alpha < \omega_1$. (ii) and (iii) are replaced respectively by (2) and (3):

(2) For every $\aleph_1$-test $(X_\alpha : \alpha < \omega_1)$, the set
$S = \{\alpha < \omega_1 : \exists i \in I \ (F(i) \subseteq X_\alpha \wedge \forall \beta < \alpha \ (F(i) \not\subseteq X_\beta))\}$ is not stationary.

(3) For every $\aleph_1$-test $(X_\alpha : \alpha < \omega_1)$, the set
$T = \{\alpha < \omega_1 : \exists \beta > \alpha \ (F_{X_\alpha}^{X_\beta}$ has no marriage$)\}$ is not stationary.

We want to show that the family in Example VI does not satisfy (2) and the family in Example VII does not satisfy (3). For this reason we define, for each $\alpha < \omega_1$, $X_\alpha = \omega_1 + \alpha$, $A_\alpha = (\omega_1 + \alpha) \setminus \omega_1$, and $K_\alpha = \omega_1$. Then K_α is critical in $F \setminus A_\alpha$, $X_\alpha = F(K_\alpha) \cup A_\alpha$, and $|A_\alpha| \leq \aleph_1$. Consequently $(X_\alpha : \alpha < \omega_1)$ is an ω_1-test. For the corresponding sets S and T in (2) and (3) we have $L(\omega_1) \subseteq S$ for VI and $L(\omega_1) \subseteq T$ for VII, hence S and T are stationary.

μ-tests are investigated in Section II.3. There it will be shown that (2) and (3) are satisfied for F if F has a marriage. Conversely we shall prove in Section II.4 that a family F with $\|F\| \leq \aleph_1$ possesses a marriage if (1), (2), and (3) are satisfied.

Especially this holds for families F with $|F(I)| \leq \aleph_1$, since in this case $\|F\| \leq \aleph_1$. In other words, we essentially obtain Shelah's condition, stated in C, modulo critical sets. Further $\|F_{X_\alpha}^{X_\beta}\| \leq \aleph_0$ for each ω_1-test $(X_\alpha : \alpha < \omega_1)$ and any $\alpha, \beta < \omega_1$, so that we can replace the statement "$F_{X_\alpha}^{X_\beta}$ has no marriage" in (3) by "$F_{X_\alpha}^{X_\beta}$ does not satisfy (1)".

E. Finally we discuss the general case. Let F be a family, and let $\lambda = \|F\|$. If F has a marriage, then the following conditions must be fulfilled:

(1) For each set A and each K critical in F_A, $|\{i \in I \setminus K : F(i) \subseteq F(K)\}| \leq |A|$.

(2) If $\mu \leq \lambda$ is an uncountable regular cardinal and $(X_\alpha : \alpha < \mu)$ is a μ-test, then there exists a closed unbounded subset C of μ such that, for each $\alpha \in C$ and each $i \in I$ with $F(i) \subseteq X_\alpha$, there is a $\beta < \alpha$ such that $F(i) \subseteq X_\beta$.

(3) If $\mu \leq \lambda$ is an uncountable regular cardinal and $(X_\alpha : \alpha < \mu)$ is a μ-test, then there exists a closed unbounded subset C of μ such that, for any $\alpha \in C$ and any $\beta \geq \alpha$, the family $F_{X_\alpha}^{X_\beta}$ has a marriage.

(4) If $\|F^X\| < \lambda$, then F^X has a marriage.

In Section II.4 will be shown that these conditions are also sufficient for the existence of a marriage. The notion of a marriage in the formulation of (3) and (4) can be eliminated by an inductive definition, since the corresponding families have an essential size less than λ.

A similar criterion - it was the first criterion for arbitrary families - was given by Aharoni, Nash-Williams, and Shelah in [ANS1]. They define the notion of a μ-obstruction and show that a family has a marriage if and only if it possesses no μ-obstruction - that means that it does not contain one of a set of forbidden substructures. In Section II.5 we present a direct proof for the equivalence of both criteria. As a consequence we obtain Shelah's Compactness Theorem for singular cardinals.

A further consequence of the Aharoni-Nash-Williams-Shelah-criterion is the following theorem of Aharoni. If $F = (F(i) : i \in I)$ is a family, let G be the family $(G(w) : w \in F(I))$, defined by $G(w) = \{i \in I : w \in F(i)\}$ and let, for any $X \subseteq F(I)$, $D(X) = \{i \in I : F(i) \subseteq X\}$. Then F has a marriage if and only if there exists a set $X \subseteq F(I)$ such that $F \upharpoonright D(X) = (F(i) : i \in D(X))$ has no marriage and $G \upharpoonright X = \{G(w) : w \in X\}$ has a marriage g with $g[X] \subseteq D(X)$. The proof is given in Section II.7. An important consequence is the following theorem, called König's Duality Theorem in [A3]: If $\Gamma = (M,W,E)$ is a bipartite graph, then there exist a set $\mathcal{O}$ of vertices and a set f of pairwise disjoint edges such that every edge of Γ contains a vertex of $\mathcal{O}$ and $\mathcal{O}$ consists of a choice of precisely one vertex from each edge of f. With the help of this theorem one can prove a strong version of Menger's Theorem for graphs containing no infinite path. These applications of marriage criteria are proved in Section II.6.

Chapter I

SET THEORETIC FOUNDATIONS

This chapter has a double aim. First it is to present the most important definitions and notations, Fodor's Theorem, and a choice lemma for stationary sets for those readers who are familiar with the techniques of set theory. On the other hand it should enable those readers who are applying transfinite methods of set theory rather scarcely to study the results of this book without using any other text on set theory. It would be regrettable if the acquaintance with the beautiful theorems of the following chapters was thwarted by the mistrust of notions such as regular cardinal, transfinite induction and so on. Therefore we advise the readers of the first kind to read Sections 1 and 6 and the readers of the second kind to start their studies with Section 2.

§1. Definitions and notations

If A and B are sets and A is a finite subset of B, then we write $A \subset\subset B$.

A *relation* is a set of ordered pairs. If R is a relation and A is a set, then $R \restriction A$, the *restriction* of R to A, denotes $\{(x,y) : (x,y) \in R$ and $x \in A\}$. $R^{-1} = \{(x,y) : (y,x) \in R\}$ is the *inverse of R*. The *domain* of R is $\mathrm{dom}\, R = \{x : (x,y) \in R$ for some $y\}$ and the *range* of R is $\mathrm{rng}\, R = \{y : (x,y) \in R$ for some $x\}$. $R[A]$, the *image of A under R*, denotes $\{y : (x,y) \in R$ for some $x \in A\}$. We write $R<a>$ for $R[\{a\}]$. We say that a relation F is a *function* if $(x,y) \in F$ and $(x,z) \in F$ always imply $y = z$. If F is a function, we write $F(a)$ for the unique b with $(a,b) \in F$; F is called *injective* if F^{-1} is a function.

If A and B are sets, then $f : A \to B$ means that f is a function with $\mathrm{dom}\, f = A$ and $\mathrm{rng}\, f \subseteq B$. We say that f is a function from A into B. $f : A \to B$ is called *surjective* if $f[A] = B$, and *bijective* if it is surjective and injective. A *family* is a function. As usual we use another notation and say that a *family* F is a function $F : I \to B$ from a so-called index set I into a set B, and we write it in the form $F = (F(i) : i \in I)$ or $F = (A_i : i \in I)$ with $A_i = F(i)$.

A choice function for a family $F = (F(i) : i \in I)$ is a function $f : I \to \bigcup \operatorname{rng} F$ such that $f(i) \in F(i)$ for every $i \in I$. We call an injective choice function for F a marriage of F. If $J \subseteq I$, then a marriage of $F \upharpoonright J$ is also called a matching of J.
Lower case Greek letters always denote ordinal numbers - their theory is developed in Section 3. As usual, every ordinal $\alpha = \{\beta : \beta < \alpha\}$ is the set of all smaller ordinals. If α is of the form $\beta \cup \{\beta\}$, then we call α a successor ordinal and write $\beta+1$ for $\beta \cup \{\beta\}$. If $\alpha \neq 0$ is not a successor ordinal, we call α a limit ordinal and write $\operatorname{Lim}(\alpha)$. $\alpha+\beta$ denotes the ordinal sum of α and β. Since we denote cardinals with $\varkappa, \lambda, \mu$ etc., there will be no confusion with the cardinal sum $\varkappa+\lambda$.
The cardinal number or cardinality of a set A, denoted by $|A|$, is the least ordinal $\varkappa$ such that there is a bijective function $f : \varkappa \to A$. In particular, a cardinal number or cardinal is an ordinal $\varkappa$ such that $|\varkappa| = \varkappa$. The smallest infinite cardinal is denoted by $\aleph_0$ or ω_0 or simply by ω, the smallest uncountable cardinal (or ordinal) by $\aleph_1$ or ω_1. As usual, we identify $\aleph_\alpha$ and ω_α. If $\varkappa$ is a cardinal, we denote by $\varkappa^+$ the least cardinal which is greater than $\varkappa$, and by $[A]^{<\varkappa}$ the set $\{x \subseteq A : |x| < \varkappa\}$.
A sequence is a function $f : \alpha \to B$ from an ordinal into a set. In most cases we prefere the notation $(a_\beta : \beta < \alpha)$ with $a_\beta = f(\beta)$. An enumeration of a set A is an injective sequence $(a_\beta : \beta < \alpha)$ such that $\{a_\beta : \beta < \alpha\} = A$. If $(a_\gamma : \gamma < \alpha)$ and $(b_\gamma : \gamma < \beta)$ are two sequences, we denote by $(a_\gamma : \gamma < \alpha) * (b_\gamma : \gamma < \beta)$ the sequence $(c_\gamma : \gamma < \alpha+\beta)$ such that $c_\gamma = a_\gamma$ if $0 \leq \gamma < \alpha$ and $c_\gamma = b_\gamma$ if $\alpha \leq \gamma < \alpha+\beta$.
If α is an ordinal and A is a subset of α, then A is said to be cofinal or unbounded in α if $A \setminus \beta \neq \emptyset$ for each $\beta < \alpha$. If α is a limit ordinal, the cofinality of α, denoted by $\operatorname{cf}(\alpha)$, is the smallest ordinal μ such that there is a function $f : \mu \to \alpha$ with $\operatorname{rng} f$ cofinal in α. We say that the cardinal number $\varkappa$ is regular if $\operatorname{cf}(\varkappa) = \varkappa$, and that $\varkappa$ is singular if $\operatorname{cf}(\varkappa) < \varkappa$. A subset $C \subseteq \varkappa$ is called closed (in $\varkappa$) if $\sup D \in C \cup \{\varkappa\}$ for each non-empty subset D of C. If $C \subseteq \varkappa$ is closed and unbounded in $\varkappa$, we call C a club (in $\varkappa$). If $\varkappa > \aleph_0$ is regular and $S \subseteq \varkappa$, we say that S is stationary (in $\varkappa$) if $S \cap C \neq \emptyset$ for each club C in $\varkappa$. A function $f : D \to \varkappa$ from a subset $D \subseteq \varkappa$ into $\varkappa$ is called regressive if $f(\alpha) < \alpha$ for each $\alpha \in D \setminus \{0\}$.

The following facts on regular cardinals are very important - we prove them in Section 6. Let $\varkappa > \aleph_0$ be a regular cardinal. Then the following holds:

1. If $(C_\alpha : \alpha<\gamma)$ is a sequence of less than κ clubs in κ, then $D = \bigcap\{C_\alpha : \alpha<\gamma\}$ is a club in κ. This implies that the union of less than κ sets which are not stationary in κ is not stationary in κ.

2. (Fodor's Theorem)
If $S \subseteq \kappa$ is stationary and $f : S \to \kappa$ is a regressive function, then there are a $\gamma<\kappa$ and a stationary set $S' \subseteq S$ such that $f(\alpha) = \gamma$ for every $\alpha \in S'$.

3. (Choice Lemma)
If $\vartheta = (D_\alpha : \alpha < \kappa)$ is a family of subsets of κ which are not stationary in κ and if $S = \bigcup\{D_\alpha : \alpha < \kappa\}$ is stationary in κ, then there are a subfamily $\vartheta' = (D_{\alpha_\nu} : \nu < \kappa)$ of ϑ and an injective choice function g for ϑ' such that rng g is stationary in κ.

§2. The axioms of ZFC

At the beginning of this century the discovery of so-called antinomies in the set theory of G.Cantor led to a foundational crisis in mathematics. The way out which was adopted by most of the mathematicians is the axiomatization of set theory; and one of the most important axiom systems is that of Zermelo and Fraenkel with the Axiom of Choice, denoted by ZFC. For some mathematical areas systems using variables which are ranging over classes are preferable, but without too great a risk we can say that the mathematician usually is working in a universe whose objects are sets, and these sets satisfy the axioms of ZFC, which we are listing now ($a \in A$ means that a is an element of A).

(Ax 1) Axiom of Extensionality
If the sets A and B have the same elements, then A = B.

(Ax 2) Axiom of the Empty Set
There is a set with no elements.
By (Ax 1), this set uniquely determined. We denote it by $\emptyset$ or 0.

(Ax 3) Axiom of Separation
If φ is a property (with parameter p), then for any sets A and p there exists a set B whose elements are just those $a \in A$ that have the property φ.

(Ax 4) Axiom of Pairing

For any sets a and b, there is a set c which has just the elements a and b.

We write $c = \{a,b\}$. $\{a,a\}$ is denoted by $\{a\}$. The <u>ordered pair</u> (a,b) is introduced by $(a,b) = \{\{a\},\{a,b\}\}$. We only have to keep in mind the property $(a_1,b_1) = (a_2,b_2)$ if and only if $a_1 = a_2$ and $b_1 = b_2$.

(Ax 5) Axiom of Union

For any set A, there is a set B whose elements are just the elements of any element of A.

We write $B = \cup A$. If $A = \{c,d\}$ we write $B = c \cup d$. For $n \geq 1$ let $\{a_0,\ldots,a_n\} = \{a_0,\ldots,a_{n-1}\} \cup \{a_n\}$.

(Ax 6) Axiom of Power Set

For any set A, there is a set B whose elements are just the subsets of A.

We denote B by $\mathcal{P}(A)$.

(Ax 7) Axiom of Infinity

There is an infinite set, or, more precisely: There exists a set A with $\emptyset \in A$, such that, for each $a \in A$, the set $a \cup \{a\}$ is an element of A.

(Ax 8) Axiom Schema of Replacement

If the property $\varphi(x,y,p)$ is functional, i.e. $\varphi(x,y_1,p)$ and $\varphi(x,y_2,p)$ always imply $y_1 = y_2$, then for each set A there exists a set B whose elements are just those b for which $\varphi(a,b,p)$ holds for some $a \in A$.

Intuitively this means of course: If we restrict a function to a set, then the range of the restriction is a set.

(Ax 9) Axiom of Foundation

If the set A is not empty, then A has an $\in$-minimal element, i.e. there exists a $b \in A$ such that b and A have no common element.

An easy consequence of (Ax 9) is the fact that there is no set a with $a \in a$: apply (Ax 9) to $A = \{a\}$. More general, there are no sets $a_1,\ldots,a_n$ such that $a_i \in a_{i+1}$ for all $i < n$ and $a_n \in a_1$: apply (Ax 9) to $A = \{a_1,\ldots,a_n\}$.

(Ax 10) Axiom of Choice

If A is a set of non-empty sets, then there is a function f such that $\mathrm{dom}\, f = A$ and $f(a) \in a$ for every $a \in A$.

We denote the system with the axioms (Ax 1) - (Ax 9) by ZF and the system with the axioms (Ax 1) - (Ax 10) by ZFC.

To make the undefined notion "property" precise, axiomatic set theory is developed in the framework of the first order predicate calculus. Apart from the equality predicate $=$, the language of set theory consists of the binary predicate $\in$, the <u>membership relation</u>. Starting with so-called <u>atomic formulas</u> $x \in y$, $x = y$, we build up the <u>formulas</u> of set theory by means of <u>connectives</u> $\wedge$(conjunction), $\vee$(disjunction), $\neg$(negation), $\rightarrow$(implication), $\leftrightarrow$(equivalence) and quantifiers $\forall$, $\exists$. That means: if φ and Ψ are formulas, then $(\varphi \wedge \Psi)$, $(\varphi \vee \Psi)$, $\neg\varphi$, $(\varphi \rightarrow \Psi)$, $(\varphi \leftrightarrow \Psi)$, $\forall x \varphi$ and $\exists x \varphi$ are formulas. Now, a <u>property</u> is a formula of set theory. As usual, $\forall x \in y \varphi$ denotes $\forall x (x \in y \rightarrow \varphi)$ and $\exists x \in y \varphi$ denotes $\exists x (x \in y \wedge \varphi)$. Of course we shall use in practice further defined predicates, operations, and constants, but these can all be eliminated.

Very advantageous is the use of <u>classes</u>. If $\varphi(x,p)$ is a formula with the free variable x, we call $\{x : \varphi(x,p)\}$ a <u>class</u>. Elements of $X = \{x : \varphi(x,p)\}$ are, as usual, just those sets a for which $\varphi(a,p)$ holds. That means $a \in \{x : \varphi(x,p)\} \leftrightarrow \varphi(a,p)$. At the same time this is the rule by which the symbols $\{x : \varphi(x,p)\}$ can be eliminated. If $Y = \{x : \Psi(x,q)\}$ is a further class, we write $X = Y$ for $\forall x (\varphi(x,p) \leftrightarrow \Psi(x,q))$ and $X \in Y$ for $\exists x (\forall y (y \in x \leftrightarrow \varphi(y,p)) \wedge \Psi(x,q))$ and so on. Every set A can be regarded as a class since $A = \{x : x \in A\}$. Especially $\{x : \varphi(x,p)\}$ is a set if there is a set A with $A = \{x : \varphi(x,p)\}$. If $\{x : \varphi(x,p)\} \neq A$ for every set A, then we call $\{x : \varphi(x,p)\}$ a <u>proper class</u>. We have seen that the use of classes is always eliminable.

(Ax 3) guarantees that, for each set A, the class $\{x : x \in A \wedge \varphi(x,p)\}$, which we also denote by $\{x \in A : \varphi(x,p)\}$, is a set: "Every subclass of a set is a set". We also call a class of ordered pairs, written $\{(x,y) : \varphi(x,y,p)\}$, a <u>relation</u>, and generalize the definitions of Section 1 in an obvious way to classes; especially $R \restriction A$, R^{-1}, $\mathrm{dom}\, R$, $\mathrm{rng}\, R$, $R[A]$, $R\langle a\rangle$, "R is a function", "R is injective" can be defined in the same way. Now we can reformulate Axiom 9 in a more readable version.

(Ax 9) If A is a set and $F = \{(x,y) : \varphi(x,y,p)\}$ is a function, then $F[A] = \{y : \exists x (x \in A \wedge \varphi(x,y,p))\}$ is a set.

Starting with our axioms, we can develop the usual calculus of boolean operations, of relations, and of functions, which is familiar to the reader. The notations can be found in Section 1.

§ 3. Ordinal numbers

Definition. Let X be a class, and let $R \subseteq X \times X$ be a relation. We say that R is a *partial ordering* of X if $\neg(a,a) \in R$ and $((a,b) \in R \wedge (b,c) \in R) \rightarrow (a,c) \in R$ for any $a,b,c \in X$. R is a *linear ordering* of X if R is a partial ordering of X and, for any $a,b \in X$, we have $(a,b) \in R \vee a = b \vee (b,a) \in R$. A linear ordering R of X is called a *well-ordering* of X if every non-empty subset B of X has an *R-smallest element*, i.e. there exists a $b \in B$ such that $(b,z) \in R$ for every $z \in B \setminus \{b\}$, and if for every $a \in X$ the class $\{b \in X : (b,a) \in R\}$ is a set. If X is a set, this last condition is superfluous; in this case we call (X,R) a well ordering, too. We usually write $a < b$ for $(a,b) \in R$ and $a \leq b$ for $(a,b) \in R \vee a = b$.

We show in this section that every ordinal number is well ordered by $\in$; in Section 4 it is shown that every well ordering $(A,<)$ is isomorphic to $(\alpha,\in)$ for some unique ordinal α. This means that the study of well orderings can be replaced by the study of ordinals. The usual induction and recursion principles for the set of natural numbers can be generalized to well orderings, hence to ordinals (see Section 4), and this yields the fundamental tools in infinite set theory. Finally the axiom of choice implies that every set can be well ordered (see Section 5); especially this allows us to define its cardinality.

Definition. A set x is called *transitive* if every element of x is a subset of x, i.e. $z \in y$ and $y \in x$ imply $z \in x$ for any y,z. We say that x is an *ordinal number* or an *ordinal* if x is transitive and every element of x is transitive. *On* $:= \{x : x$ is an ordinal$\}$ is the class of ordinals. Lower case Greek letters always denote ordinals. We write $\sigma < \tau$ for $\sigma \in \tau$ and $\sigma \leq \tau$ for $\sigma \in \tau \vee \sigma = \tau$.

Lemma 3.1.
(a) Every element of an ordinal is an ordinal.
(b) On is not a set.

Proof. For (a), let $x \in \sigma$. Then x is transitive by definition, and $x \subseteq \sigma$. Therefore every element of x is transitive, and x is an ordinal.
For the proof of (b), assume that $On = A$ is a set. By (a), A is an ordinal, and therefore $A \in A$, which contradicts (Ax 9).

Lemma 3.2. $<$ is a partial ordering on On.

Proof. By (Ax 9), we have $\neg(\sigma < \sigma)$. If $\sigma < \tau$ and $\tau < \rho$, then $\sigma < \rho$, since ρ is transitive.

Lemma 3.3. Every non-empty class of ordinals has an $\in$-minimal element.

Proof. Let $D \subseteq On$, $D \neq \emptyset$, and choose $\sigma \in D$. If $\sigma \cap D = \emptyset$, we are done. Otherwise $z = \{\rho < \sigma : \rho \in D\}$ is a non-empty set by (Ax 3), which has an $\in$-minimal element τ by (Ax 9). If $\alpha < \tau$, then $\alpha < \sigma$, but $\alpha \notin z$. Thus $\alpha \notin D$, which shows that τ is an $\in$-minimal element of D.

Lemma 3.4. $\sigma < \tau$ or $\sigma = \tau$ or $\tau < \sigma$.

Proof. We assume that there are ordinals σ and τ for which the lemma fails. Then $x_1 = \{\sigma : \exists \tau (\neg \sigma < \tau \wedge \sigma \neq \tau \wedge \neg \tau < \sigma)\} \neq \emptyset$. By Lemma 3.3, there is an $\in$-minimal element $\sigma^* \in x_1$. Let $x_2 = \{\tau : \neg \sigma^* < \tau \wedge \sigma^* \neq \tau \wedge \neg \tau < \sigma^*\}$, and choose τ^* as an $\in$-minimal element of x_2. We show that $\sigma^* = \tau^*$, and this contradiction proves the lemma. Let $\rho \in \tau^*$ (see 3.1(a)). Then $\rho \notin x_2$ by the choice of τ^*, hence $\rho < \sigma^*$ or $\sigma^* = \rho$ or $\sigma^* < \rho$. $\sigma^* \leq \rho$ implies $\sigma^* \in \tau^*$. Since $\tau^* \in x_2$, we have $\rho < \sigma^*$ and so $\tau^* \subseteq \sigma^*$. Analogously we get from $\rho \in \sigma^*$ that $\rho \notin x_1$ and $\rho \in \tau^*$; consequently we have $\sigma^* = \tau^*$.

Corollary 3.5.
(a) On is well-ordered by $\in$.
(b) Every ordinal is well-ordered by $\in$.
(c) $\sigma \leq \tau$ if and only if $\sigma \subseteq \tau$.

<u>Lemma 3.6.</u>

(a) If A is a set of ordinals, then $\cup A$ is an ordinal and $\cup A$ is the supremum of A.

(b) If $A \neq \emptyset$ is a class of ordinals, then $\cap A = \{x : \forall y \in A(x \in y)\}$ is the minimum of A.

<u>Proof.</u> For (a), it is easy to show that $\cup A$ is an ordinal. If $\tau \in A$, then $\tau \subseteq \cup A$ and therefore $\tau \leq \cup A$ by 3.5. Therefore $\cup A$ is an upper bound of A. If $\sigma < \cup A$, then $\sigma \in \rho$ for some $\rho \in A$, hence σ is not an upper bound of A. This proves (a). For (b), it is easy to see that the $\in$-minimal element of A (see 3.3), which is the minimum of A by 3.4, equals $\cap A$.

<u>Definition.</u> $\sigma \cup \{\sigma\}$, the <u>successor</u> of σ, is denoted by $S(\sigma)$ or by $\sigma + 1$. We put $0 = \emptyset$, $1 = S(\emptyset)$, $2 = S(1)$ etc.. σ is called a <u>successor ordinal</u> if there exists a β with $\beta + 1 = \sigma$. σ is called a <u>limit ordinal</u>, denoted by $Lim(\sigma)$, if $\sigma \neq 0$ and σ is not a successor number.

<u>Lemma 3.7</u>

(a) $S(\sigma)$ is an ordinal and $\sigma < S(\sigma)$.

(b) $\sigma \leq \tau \leq S(\sigma)$ implies $\tau = \sigma$ or $\tau = S(\sigma)$.

(c) $Lim(\alpha) \leftrightarrow \alpha \neq 0 \wedge \cup\alpha = \alpha$.

(d) Every ordinal either equals 0 or is a successor ordinal or is a limit ordinal.

<u>Proof.</u> (a) is obvious. If $\sigma < \tau < S(\sigma)$, then $\sigma < \tau < \sigma$ or $\sigma < \tau = \sigma$, hence $\sigma < \sigma$ by 3.2, a contradiction. This proves (b). For (c), let α be a limit ordinal. Then $\alpha \neq 0$ and $\beta + 1 < \alpha$, hence $\beta \in \beta + 1 \in \alpha$, for each $\beta \in \alpha$. This implies $\alpha \subseteq \cup\alpha$. Further $\cup\alpha \subseteq \alpha$, since α is an ordinal. If $\alpha \neq 0$ is not a limit ordinal, say $\alpha = \beta + 1$, then $\cup\alpha = \beta < \alpha$, hence $\cup\alpha \neq \alpha$, which proves (c).

<u>Lemma 3.8.</u>

(a) There exists a limit ordinal.

(b) If ω is the smallest limit ordinal and $C = \{x : 0 \in x \wedge \forall z \in x(z \cup \{z\} \in x)\}$ is the class of so-called inductive sets, then $\omega = \cap C$.

Proof. First we prove (a). (Ax 7) yields an inductive set y, i.e. we have $\emptyset \in y$ and $z \cup \{z\} \in y$ for any $z \in y$. Let $x = \{z \in y : z$ is an ordinal $\wedge \forall u \in z\ (u \in y)\}$. We show that x is a limit ordinal. The set x is non-empty, since $0 \in x$, and every element of x is transitive. If $u \in z$ and $z \in x$, then $u \in x$: first we get $u \in y$ and u is an ordinal; if $v < u$, then $v < z$ and therefore $v \in y$. So x is transitive and therefore x is an ordinal. If $x = S(\sigma)$ for some σ, then $\sigma \in y$, hence $S(\sigma) \in y$, and for any $u < S(\sigma)$ we have $u \in y$; therefore $S(\sigma) \in x = S(\sigma)$, which contradicts (Ax 9). Thus we have shown that x is a limit ordinal.
Now we prove (b). Since ω is a limit ordinal, we have $0 < \omega$ and $S(\sigma) < \omega$ for any $\sigma \in \omega$. Therefore $\omega \in C$ and $\cap C \subseteq \omega$. Assume that $\omega \setminus \cap C \neq \emptyset$ and choose the smallest $\sigma \in \omega \setminus \cap C$. $\sigma \neq 0$, since $0 \in \cap C$. By the definition of ω, we have $\sigma = \tau + 1$ for some τ. Hence $\tau \in \cap C$, and therefore $\tau + 1 \in \cap C$ by the definition of C, a contradiction. So we get $\omega = \cap C$.

Corollary 3.9. (Induction principle)
If A is a subset of ω such that $0 \in A$ and $\sigma + 1 \in A$ for every $\sigma \in A$, then $A = \omega$.

Definition. The smallest limit ordinal ω is also called the set of natural numbers, each element of ω is called a natural number.

§4. Transfinite induction and transfinite recursion

Theorem 4.1. (Transfinite induction)
If $\varphi(z,p)$ is a property such that $\forall \sigma (\forall \tau < \sigma\ \varphi(\tau,p) \rightarrow \varphi(\sigma,p))$, then $\forall \sigma\ \varphi(\sigma,p)$.

Proof. We assume that there is a σ with $\neg\varphi(\sigma,p)$. By Lemma 3.3, there is a smallest σ_0 with $\neg\varphi(\sigma_0,p)$; so we have $\forall \tau < \sigma_0\ \varphi(\tau,p)$. The hypothesis yields $\varphi(\sigma_0,p)$, a contradiction.

From Theorem 4.1 we obtain the well-known induction principle, $\forall n \in \omega (\forall m < n\ \varphi(m,p) \rightarrow \varphi(n,p))$ implies $\forall n \in \omega\ \varphi(n,p)$, as a corollary. Often it is useful to apply the following version of Theorem 4.1, which uses 3.7(d):

Theorem 4.1*.
If $\varphi(z,p)$ is a property with
(i) $\varphi(0,p)$,
(ii) $\forall\sigma\,(\varphi(\sigma,p) \to \varphi(\sigma+1,p))$, and
(iii) $\forall\sigma\,((\mathrm{Lim}(\sigma) \wedge \forall\tau<\sigma\ \varphi(\tau,p)) \to \varphi(\sigma,p))$,
then $\forall\sigma\ \varphi(\sigma,p)$.

Theorem 4.2. (Transfinite recursion)
If the class G is a function, defined for every set, then there exists exactly one function F with $\mathrm{dom}\,F = On$ such that, for all $\sigma \in On$,

$$F(\sigma) = G(F \upharpoonright \sigma).$$

Proof. Take $C = \{f : f \text{ is a function} \wedge \mathrm{dom}\,f \in On \wedge \forall\tau \in \mathrm{dom}\,f\ (f(\tau) = G(f \upharpoonright \tau))\}$, and let $F = \bigcup C$.

Claim 1. F is a function.

Let $f_1, f_2 \in C$ and $\alpha \in \mathrm{dom}\,f_1 \cap \mathrm{dom}\,f_2 =: \rho$. We prove $f_1(\alpha) = f_2(\alpha)$ by transfinite induction on α. If $\sigma<\alpha$, then $\sigma \in \rho$, since $\rho \in On$, so we have $f_1(\sigma) = f_2(\sigma)$ by the inductive hypothesis. Therefore $f_1 \upharpoonright \alpha = f_2 \upharpoonright \alpha$ and $f_1(\alpha) = G(f_1 \upharpoonright \alpha) = G(f_2 \upharpoonright \alpha) = f_2(\alpha)$, hence F is a function.

Claim 2. For every $\sigma \in \mathrm{dom}\,F$ we have $F(\sigma) = G(F \upharpoonright \sigma)$.

Let $\sigma \in \mathrm{dom}\,F$ and $f \in C$ with $\sigma \in \mathrm{dom}\,f$. By Claim 1, we have $f \upharpoonright \sigma = F \upharpoonright \sigma$, and since $f \in C$, we get $F(\sigma) = f(\sigma) = G(f \upharpoonright \sigma) = G(F \upharpoonright \sigma)$.

Claim 3. $\mathrm{dom}\,F = On$.

It is clear that $\mathrm{dom}\,F \subseteq On$. We apply Theorem 4.1 and assume that $\tau \in \mathrm{dom}\,F$ for every $\tau<\sigma$. Put $h = F \upharpoonright \sigma \cup \{(\sigma, G(F \upharpoonright \sigma))\}$. h is a function with $\mathrm{dom}\,h = \sigma+1 \in On$. If $\tau<\sigma$, then $h(\tau) = F(\tau) = G(F \upharpoonright \tau) = G(h \upharpoonright \tau)$. Further we have $h(\sigma) = G(F \upharpoonright \sigma) = G(h \upharpoonright \sigma)$. So $h \in C$, and consequently $\sigma \in \mathrm{dom}\,F$. We have shown by transfinite induction that $On \subseteq \mathrm{dom}\,F$.

Claim 4: If F_1 is a function with $\mathrm{dom}\,F_1 = On$ such that $F_1(\sigma) = G(F_1 \upharpoonright \sigma)$ for every $\sigma \in On$, then $F_1 = F$.
This is easily shown by transfinite induction.

Often it is useful to apply the following version of Theorem 4.2, corresponding to Theorem 4.1*:

Theorem 4.2*.

Let G_1 and G_2 be functions which are defined for every set, and let a be a set. Then there exists exactly one function F with dom F = On such that (i) $F(0) = a$,

(ii) $F(\alpha+1) = G_1(F(\alpha))$, and

(iii) $F(\alpha) = G_2(F \upharpoonright \alpha)$ for any limit ordinal α.

Proof. Define G by

$(x,y) \in G \leftrightarrow (\neg(x \text{ is a function} \wedge \text{dom}\, x \in On) \wedge y = 0) \vee \varphi(x,y,a)$,

where $\varphi(x,y,a)$ means that x is a function

$((x = \emptyset \wedge y = a) \vee \exists\alpha(\text{dom}\, x = \alpha+1 \wedge y = G_1(x(\alpha))) \vee \exists\alpha(\alpha \text{ is a limit ordinal} \wedge \text{dom}\, x = \alpha \wedge y = G_2(x \upharpoonright \alpha)))$.

G is a function which is defined for every set, and the function F, yielded by Theorem 4.2, has the desired properties.

Remark: Theorem 4.2. is a metatheorem, since it says something about the existence of classes. A formally correct version would be the following: To every formula $\Psi(x,y)$ (corresponding to G), we can construct a formula $\varphi(x,y)$ (corresponding to F) such that in ZF the following theorem is provable: Assume that for every f there is exactly one y with $\Psi(f,y)$. Then there is for every α exactly one y with $\varphi(\alpha,y)$. Further, if f is a function with domain α such that $\varphi(\beta,f(\beta))$ for any $\beta<\alpha$, then $\varphi(\alpha,y)$ is equivalent to $\Psi(f,y)$. The proof of Theorem 4.2 corresponds to this version.

Lemma 4.3. If $f:\alpha\rightarrow\beta$ is an order isomorphism, then $\alpha = \beta$ and $f = id_\alpha$.

Proof. Let σ be the smallest ordinal with $f(\sigma) \neq \sigma$. Then $f[\sigma] = \sigma$, hence $f(\sigma)>\sigma$ and $f^{-1}(\sigma)>\sigma$. But $f^{-1}(\sigma)>\sigma$ implies $\sigma>f(\sigma)$, since f is strictly increasing. So $\sigma<\sigma$, a contradiction. We have $f = id_\alpha$ and, since f is onto, $\alpha = \beta$.

Theorem 4.4.

(a) If $(A,<)$ is a well-ordering, then there exists exactly one ordinal α and one function f such that f is an isomorphism from $(\alpha,\in)$ onto $(A,<)$.

(b) If C is a proper class which is well-ordered by $<$, then C is order isomorphic to On, well-ordered by $\in$.

Proof. For the proof of (a), let $b \neq \emptyset$ be a set with $b \not\in A$ and define G by

$$(x,y) \in G \leftrightarrow (\neg(x \text{ is a function} \wedge \operatorname{dom} x \in On) \wedge y = 0) \vee \varphi(x,y)$$

where $\varphi(x,y)$ means that x is a function with $\operatorname{dom} x \in On$ and $\operatorname{rng} x \subseteq A \cup \{b\}$ and $y = \min(A \setminus \operatorname{rng} x)$ if $A \setminus \operatorname{rng} x \neq \emptyset$, $y = b$ otherwise. Let F be the function according to Theorem 4.2, and let $\alpha = \min\{\sigma : F(\sigma) = b\}$. Put $f = F \upharpoonright \alpha$. By (Ax 8), f is a set. f is injective, since, for all $\beta < \alpha$, $f(\beta)$ is the $<$-smallest element of $A \setminus \operatorname{rng} f \upharpoonright \beta$. Further we have $f[\alpha] = A$ by the choice of α. By transfinite induction, one can easily show that for every $\gamma \leq \alpha$ the set $A_\gamma = \operatorname{rng} f \upharpoonright \gamma$ is an initial segment of A (i.e. $y < z$ and $z \in A_\gamma$ implies $y \in A_\gamma$ for every $y \in A$), and that $f \upharpoonright \gamma : \gamma \to A_\gamma$ is an isomorphism. Lemma 3.3 implies that α and f are uniquely determined. - The proof of (b) runs analogously. We don't need the set b, since always $C \setminus \operatorname{rng} x \neq \emptyset$ by (Ax 8) and because C is not a set. $F[On]$ is an initial segment of C, On is not a set, and F is injective. Therefore, by (Ax 8), $F[On]$ is not a set. Consequently there is no $a \in C \setminus F[On]$ by the definition of a well-ordered class.

If α is an ordinal, then, by Theorem 4.2*, there exists a function f_α such that $f_\alpha(0) = \alpha$, $f_\alpha(\beta+1) = f_\alpha(\beta)+1$ for any β, and $f_\alpha(\gamma) = \sup\{f_\alpha(\beta) : \beta < \gamma\}$ for limit ordinals γ. We write $\alpha + \beta$ for $f_\alpha(\beta)$ and call it the ordinal sum of α and β. A more natural approach to $\alpha+\beta$ is the following: If $(A_1,<_1)$ and $(A_2,<_2)$ are two well orderings with $A_1 \cap A_2 = \emptyset$, then define the relation $<$ on $A_1 \cup A_2$ by $< \; = \; <_1 \cup <_2 \cup A_1 \times A_2$. It is easy to see that $<$ is a well ordering of $A_1 \cup A_2$. One can show that $(A_1 \cup A_2, <)$ is isomorphic to $(\alpha+\beta, \in)$, if $(A_1,<_1)$ is isomorphic to $(\alpha,\in)$ and $(A_2,<_2)$ to $(\beta,\in)$.

In a similiar way one can introduce multiplication and exponentiation of ordinals.

§5. Cardinal numbers

Theorem 5.1.

In the axiom system ZF, the following propositions are equivalent:

(i) The axiom of choice.

(ii) The well-ordering principle: For any set A, there exists a well-ordering of A.

(iii) Zorn's Lemma: If $(P,<)$ is a non-empty partial ordering and P is a set such that every subset of P which is linearly ordered by $<$ has an upper bound in P, then P has a maximal element.

Proof. (ii)$\to$(i): Let A be set of non-empty sets, and let $<$ be a well ordering of $\bigcup A$. Take

$f = \{(x,y) : x\in A$ and y is the $<$-smallest element of $x\}$.

Obviously f is a choice function for A.

(i)$\to$(iii): Let g be a choice function for $\mathcal{P}(P)\setminus\{\emptyset\}$ and choose $b\notin P$. We apply Theorem 4.2, but we shall argue informally from now on; it is not difficult to give a precise definition of the function G used in the theorem. Let $a_0\in P$ and let a_τ be defined for $\tau<\sigma$. Put $K_\sigma = \{a_\tau : \tau<\sigma\}$ and let, for any $Q\subseteq P$, $s(Q)$ be the set of upper bounds of Q in $P\setminus Q$. If $s(K_\sigma) = \emptyset$, define $a_\sigma = b$, otherwise define $a_\sigma = g(s(K_\sigma))$. There exists a smallest α with $a_\alpha = b$: otherwise we would get an injection h of On to P. h[On] is a set as a subset of P by (Ax 8) and the remark to Theorem 4.2, and so $h^{-1}[h[On]] = On$ is a set by (Ax 8), which contradicts Lemma 3.1. By assumption, the set K_α has a greatest element d, since it has an upper bound in P, and this is a maximal element of P.

(iii)$\to$(ii): Let $H = \{f : \exists\sigma(f:\sigma\to A$ is injective$)\}$ and let $<$ be defined by $f<g \Leftrightarrow f\subsetneq g$ for $f,g\in H$. It is clear that $<$ is a partial ordering of H, and if $K\subseteq H$ is linearly ordered by $<$, then $\bigcup K$ is an upper bound of K in H. Especially we have $\emptyset\in H$. To apply (iii), we have to show that H is a set. If $f\in H$, then f induces a well ordering $R_f = \{(a,b) : a,b\in \text{rng}\, f$ and $f^{-1}(a)<f^{-1}(b)\}$ of $\text{rng}\, f\subseteq A$ - of course this motivates the definition of H. The function $F = \{(f,(\text{rng}\, f,R_f)) : f\in H\}$ is injective by Theorem 4.4, since $(\text{rng}\, f,R_f)$ is isomorphic to exactly one ordinal. We have $\text{rng}\, F\subseteq \mathcal{P}(A)\times\mathcal{P}(A\times A)$, so rng F is a set, and consequently $H = \text{rng}\, F^{-1}$ is a set by (Ax 8). By Zorn's Lemma, there exists a maximal element h of H. Any $x\in A\setminus \text{rng}\, h$ would yield the proper extension $h\cup\{(\text{dom}\, h,x)\}$ of h, so we have $\text{rng}\, h = A$, and R_h is a well-ordering of A.

Definition. If A is a set, then the cardinal number or cardinality of A, denoted by $|A|$, is the least ordinal $\varkappa$ such that there exists a bijective function $f:\varkappa\to A$. In particular, a cardinal number or a cardinal is an ordinal $\varkappa$ such that $|\varkappa| = \varkappa$. For every set A, the ordinal $|A|$ exists. This follows from the axiom of choice, since every set has a well-ordering (Theorem 5.1) and since every well-ordered set is isomorphic to an ordinal (Theorem 4.4). Obviously every cardinal number of a set is a cardinal number, and $|\alpha| \leq \alpha$ for every ordinal α. In the sequel, $\varkappa,\lambda,\mu$ always denote cardinals.

Theorem 5.2. (Schröder-Bernstein)
Let A and B be sets such that there is an injection $f:A\to B$ and an injection $g:B\to A$. Then there exists a bijection $h:A\to B$ with $h\subseteq f\cup g^{-1}$.

Proof. We prove the theorem without the use of (Ax 10) and use the technique of alternating paths, which will be very useful in the following chapters. For any $x\in B\setminus \operatorname{rng} f$, we define a sequence $(z_n^x : n\in\omega)$ by $z_0^x = x$ and $z_{n+1}^x = f(g(z_n^x))$. Since f and g are injective, we have, for any $x,y\in B\setminus \operatorname{rng} f$ with $x \neq y$, that $\{z_n^x : n\in\omega\}\cap\{z_n^y : n\in\omega\} = \emptyset$. Therefore the following function h is the desired bijection:

$$h = (f \setminus \bigcup\{\{(g(z_n^x), f(g(z_n^x))) : n\in\omega\} : x\in B\setminus \operatorname{rng} f\}$$
$$\cup \bigcup\{\{(g(z_n^x), z_n^x) : n\in\omega\} : x\in B\setminus \operatorname{rng} f\}.$$

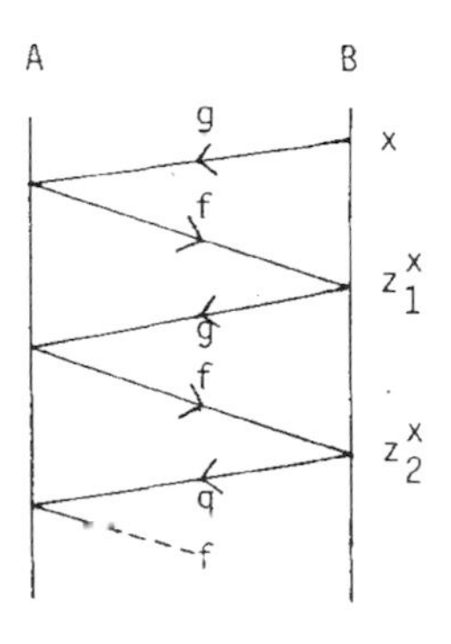

Corollary 5.3.
(a) $|A| \leq |B|$ if and only if there is an injection from A into B.
(b) $|A| = |B|$ if and only if there is a bijection from A onto B.

Lemma 5.4. If K is a set of cardinals, then $\sup K$ is a cardinal.

Proof. If we show that the ordinal $|\sup K|$ is an upper bound of K and consequently $\sup K \leq |\sup K|$, we are done, since $|\sup K| \leq \sup K$. Let $f: \sup K \to |\sup K|$ be a bijection. If $\varkappa \in K$, then $\varkappa \subseteq \bigcup K = \sup K$; hence $f | \varkappa$ is an injection from $\varkappa$ into $|\sup K|$, and by Corollary 5.3 we have $\varkappa \leq |\sup K|$.

Definition. If $\varkappa$ and λ are cardinals, we define their (cardinal) sum, their product, and the power $\varkappa^\lambda$ by
$\varkappa + \lambda = |(\varkappa \times \{0\}) \cup (\lambda \times \{1\})|$, $\varkappa \cdot \lambda = |\varkappa \times \lambda|$, and $\varkappa^\lambda = |\{f : f : \lambda \to \varkappa\}|$.

If $|A| = \varkappa$ and $|B| = \lambda$, it is easy to see that $\varkappa + \lambda = |A \times \{0\} \cup B \times \{1\}|$, $\varkappa \cdot \lambda = |A \times B|$, and $\varkappa^\lambda = |\{f : f : B \to A\}|$.

Lemma 5.5.
If $\varkappa$ is a cardinal, then
(a) $|\mathcal{P}(\varkappa)| = 2^\varkappa$, (b) $\varkappa < 2^\varkappa$.

Proof. For any $x \in \mathcal{P}(\varkappa)$, let $f(x)$ be the characteristic function of x. $f : \mathcal{P}(\varkappa) \to \{g : g : \varkappa \to \{0,1\}\}$ is bijective, and this proves (a). For the proof of (b), we note first that $\varkappa \leq |\mathcal{P}(\varkappa)|$. We show that there is no surjective $f : \varkappa \to \mathcal{P}(\varkappa)$, and so $\varkappa \neq |\mathcal{P}(\varkappa)|$. Let $f : \varkappa \to \mathcal{P}(\varkappa)$ be a function and define $y = \{\alpha < \varkappa : \alpha \notin f(\alpha)\}$. If $y = f(\sigma)$ for some $\sigma < \varkappa$, we have $\sigma \in y$ if and only if $\sigma \notin y$. This shows that f is not surjective.

Corollary 5.6. $Cn = \{\alpha : \alpha \text{ is a cardinal}\}$ is a proper class. Especially $Cn \setminus \omega$ is a proper class.

Remark. The proof of Lemma 5.5 uses the Axiom of Choice. But we can show in ZF that for every cardinal there exists a greater cardinal. If $\varkappa$ is an infinite cardinal, consider the class
$A = \{\sigma : \exists f (f : \sigma \to \varkappa \text{ is injective})\}$. In the proof of Theorem 5.1 we have shown in ZF that $H = \{f : \exists \sigma (f : \sigma \to \varkappa \text{ is injective})\}$ is a set. Consequently $A = \{\operatorname{dom} f : f \in H\}$ is a set, and the least $\alpha \in On \setminus A$ is a cardinal greater than $\varkappa$.

Definition. A set A is called finite if $|A| = n$ for some $n \in \omega$. Otherwise A is called infinite. We say that A is countable if $|A| \leq \aleph_0$. If $\varkappa$ is a cardinal, we denote by $\varkappa^+$ the least cardinal which is greater than $\varkappa$. By Theorem 4.4, Lemma 5.4, and Corollary 5.6, there exists a function $\aleph$ which yields an order preserving isomorphism from On onto $Cn \setminus \omega$. We write $\aleph_\alpha$ or ω_α for $\aleph(\alpha)$.

Lemma 5.7.
(a) If $\varkappa$ is an infinite cardinal, then there exists an α with $\varkappa = \aleph_\alpha$.
(b) If $\sigma < \tau$, then $\aleph_\sigma < \aleph_\tau$.
(c) $\aleph_0 = \omega$, $\aleph_{\alpha+1} = \aleph_\alpha^+$, and $\aleph_\gamma = \sup\{\aleph_\beta : \beta < \gamma\}$ for any limit ordinal γ.

It is well-known by the results of K. Gödel and P.J. Cohen that the axioms of ZFC do not decide whether the set of real numbers has cardinality $\aleph_1$. More general, they do not decide whether $\aleph_{\alpha+1} = |\mathcal{P}(\aleph_\alpha)|$, and so they cannot decide whether the family $F_\alpha = \{(i, \aleph_{\alpha+1}) : i \in \mathcal{P}(\aleph_\alpha)\}$ has an injective choice function.

Theorem 5.8. There exists a well-ordering $<$ of $On \times On$ such that, for any α, the set $\aleph_\alpha \times \aleph_\alpha$ is an initial segment of $On \times On$ with respect to $<$ which is isomorphic to $(\aleph_\alpha, \in)$.

Corollary 5.9. For every α, we have $\aleph_\alpha \cdot \aleph_\alpha = \aleph_\alpha$.

Proof of Theorem 5.8:
We define $<$ by $(\alpha_1, \alpha_2) < (\beta_1, \beta_2)$ if and only if $\max\{\alpha_1, \alpha_2\} < \max\{\beta_1, \beta_2\}$ or $\max\{\alpha_1, \alpha_2\} = \max\{\beta_1, \beta_2\}$ and (α_1, α_2) precedes (β_1, β_2) lexicographically, i.e. $\alpha_1 < \beta_1$ or $(\alpha_1 = \beta_1$ and $\alpha_2 < \beta_2)$. It is not difficult to prove that $<$ is a well ordering and that $\aleph_\alpha \times \aleph_\alpha$ is an initial segment of $On \times On$ with respect to $<$. To get a contradiction, we assume that there exists an α such that $(\aleph_\alpha \times \aleph_\alpha, <)$ is not isomorphic to $(\aleph_\alpha, \in)$ and choose α as the least ordinal with this property. Now there exists a unique γ and an isomorphism $f : \gamma \to \aleph_\alpha \times \aleph_\alpha$ by Theorem 4.4. Since $|\aleph_\alpha \times \aleph_\alpha| \geq \aleph_\alpha$, we have $\gamma \geq |\gamma| \geq \aleph_\alpha$. By assumption, $\gamma \neq \aleph_\alpha$ and so $\aleph_\alpha < \gamma$. Consequently $(\aleph_\alpha, \in)$ is isomorphic to a proper initial segment of $(\aleph_\alpha \times \aleph_\alpha, <)$. Especially there is a $\sigma < \aleph_\alpha$ such that $f[\aleph_\alpha] \subseteq \sigma \times \sigma$. Since σ is infinite, there exists a $\beta < \alpha$ with $|\sigma| = \aleph_\beta$; hence $\aleph_\beta \cdot \aleph_\beta = \aleph_\beta$ by

the choice of α, and we get $\aleph_\alpha = |f[\aleph_\alpha]| \leq |\sigma \times \sigma| \leq |\aleph_\beta \times \aleph_\beta| = \aleph_\beta \cdot \aleph_\beta = \aleph_\beta$, a contradiction.

Lemma 5.10.

(a) If A is infinite and $n \in \omega$, then $|A|^n = |A|$.

(b) $\aleph_\alpha \cdot \aleph_\beta = \aleph_\alpha + \aleph_\beta = \max\{\aleph_\alpha, \aleph_\beta\} = \aleph_{\max\{\alpha,\beta\}}$.

(c) $\alpha \leq \beta$ implies $\aleph_\alpha^{\aleph_\beta} = 2^{\aleph_\beta}$.

(d) If $(A_i : i \in I)$ is a family, then
$|\bigcup\{A_i : i \in I\}| \leq |I| \cdot \sup\{|A_i| : i \in I)$.

(e) If A is infinite, then $|\{y \subseteq A : y \text{ is finite}\}| = |A|$.

Proof. (a) follows at once from Corollary 5.9. For (b), let $\alpha \leq \beta$. Then $\aleph_\beta \leq \aleph_\alpha + \aleph_\beta \leq \aleph_\alpha \cdot \aleph_\beta \leq \aleph_\beta \cdot \aleph_\beta = \aleph_\beta$. For (c), we have $2^{\aleph_\beta} \leq \aleph_\alpha^{\aleph_\beta} = |\{f : f : \aleph_\beta \to \aleph_\alpha\}| \leq |\mathcal{P}(\aleph_\beta \times \aleph_\alpha)| = 2^{\aleph_\alpha \cdot \aleph_\beta} = 2^{\aleph_\beta}$ (note that every $f : \aleph_\beta \to \aleph_\alpha$ is a subset of $\aleph_\beta \times \aleph_\alpha$).

For (d), let $\varkappa = \sup\{|A_i| : i \in I\}$ and let $<$ be a well-ordering of I. For any $a \in \bigcup\{A_i : i \in I\}$, let $f(a) = (a,i)$, where i is the $<$-smallest element of I with $a \in A_i$. f is injective and $\operatorname{rng} f \subseteq \bigcup\{A_i \times \{i\} : i \in I\}$. Using Corollary 5.3 several times, we get $|\bigcup\{A_i : i \in I\}| \leq |\bigcup\{A_i \times \{i\} : i \in I\}| \leq |\bigcup\{|A_i| \times \{i\} : i \in I\}| \leq |\bigcup\{\varkappa \times \{i\} : i \in I\}| = |\varkappa \times I| = |I| \cdot \varkappa$.

For (e), choose a well-ordering $<$ of A and let $A_n = \{y \subseteq A : |y| = n\}$. For each $y \in A_n$, let $g(y)$ be the order isomorphism from n onto y. This defines an injection g from A_n into $\{f : f : n \to A\}$. By (a), we have $\sup\{|A_n| : n \in \omega\} \leq |A|$, and therefore $|\bigcup\{A_n : n \in \omega\}| \leq \omega \cdot |A| = |A|$ by (d). Clearly $|A| \leq |\bigcup\{A_n : n \in \omega\}|$, and (e) is proved.

§6. Clubs and stationary sets

Definition. If α is a limit ordinal and $A \subseteq \alpha$, then we say that A is cofinal or unbounded in α if for every $\beta < \alpha$ there exists a $\gamma \in A$ such that $\beta < \gamma$. The cofinality of α, denoted by $\operatorname{cf}(\alpha)$, is the smallest ordinal μ such that there exists a function $f : \mu \to \alpha$ with $\operatorname{rng} f$ cofinal in α. We say that an infinite cardinal number $\varkappa$ is regular if $\operatorname{cf}(\varkappa) = \varkappa$, and singular if $\operatorname{cf}(\varkappa) < \varkappa$.

Lemma 6.1.

(a) If α is a limit ordinal, then $cf(cf(\alpha)) = cf(\alpha)$, and consequently $cf(\alpha)$ is a regular cardinal.

(b) $\aleph_0$ is regular, and $\aleph_{\alpha+1}$ is regular for any α.

(c) $\aleph_\omega$ is the smallest singular cardinal.

(d) The class of singular cardinals is unbounded in On.

Proof. The proof of (a) is left to the reader. For (b), we show that, for any function $f: \beta \to \aleph_{\alpha+1}$ with $\beta < \aleph_{\alpha+1}$, we have $\sup \operatorname{rng} f < \aleph_{\alpha+1}$. Let f be such a function. Then, for any $\gamma < \beta$, we have $|f(\gamma)| < \aleph_{\alpha+1}$, hence $|f(\gamma)| \leq \aleph_\alpha$. Lemma 5.10.(d) shows that $|\sup \operatorname{rng} f| = |\bigcup f(\gamma) : \gamma < \beta\}| \leq |\beta| \cdot \aleph_\alpha = \aleph_\alpha$. Especially $\sup \operatorname{rng} f < \aleph_{\alpha+1}$, since $\aleph_{\alpha+1}$ is a cardinal. So we get $cf(\aleph_{\alpha+1}) \geq \aleph_{\alpha+1}$. Obviously $cf(\aleph_{\alpha+1}) \leq \aleph_{\alpha+1}$, and (b) is proved. For (c), we note that $\aleph_\omega = \sup\{\aleph_n : n \in \omega\}$, hence $cf(\aleph_\omega) = \omega < \aleph_\omega$, and so $\aleph_\omega$ is singular. If $\kappa < \aleph_\omega$ is an infinite cardinal, then $\kappa = \aleph_n$ for some $n \in \omega$ and, by (b), κ is regular. For (d), let $\alpha \in On$ be infinite. The ordinal sum $\alpha + \omega$ is a limit ordinal greater than α, and $\aleph_{\alpha+\omega}$ has cofinality ω, since $\aleph_{\alpha+\omega} = \sup\{\aleph_{\alpha+n} : n \in \omega\}$. Further we have $\aleph_0 = \omega \leq \aleph_\alpha < \aleph_{\alpha+\omega}$ by 5.7.

Definition. If $(X_i : i \in I)$ is a family, then we call the set

$$\Pi\{X_i : i \in I\} = \{f : I \to \bigcup\{A_i : i \in I\} : f(i) \in A_i \text{ for every } i \in I\}$$

the cartesian product of the family $(X_i : i \in I)$. Let $(\kappa_i : i \in I)$ be a family of cardinals. We define

$$\Sigma\{\kappa_i : i \in I\} = |\bigcup\{\kappa_i \times \{i\} : i \in I\}| \text{ and } \dot{\Pi}\{\kappa_i : i \in I\} = |\Pi\{\kappa_i : i \in I\}|$$

as the sum and the product of the family $(\kappa_i : i \in I)$.

If the context is clear, we omit the point at $\dot{\Pi}\{\kappa_i : i \in I\}$. The definition coincides with the usual definition for $I = \{0,1\}$. As usual one can show for every family $(X_i : i \in I)$ with the property $|X_i| = \kappa_i$ for every $i \in I$: $|\Pi\{X_i : i \in I\}| = \dot{\Pi}\{\kappa_i : i \in I\}$ and, if $X_i \cap X_j = \emptyset$ for any $i \neq j$, $|\bigcup\{X_i : i \in I\}| = \Sigma\{\kappa_i : i \in I\}$.

The prototype of an uncountable cardinal is - for our purposes - $\aleph_1$. Many proofs for $\aleph_1$ can be generalized to arbitrary regular cardinals greater than ω. So the following characterization is a natural generalization of the well-known fact that the union of countable many countable sets is countable.

Lemma 6.2. Let $\varkappa$ be an infinite cardinal. Then $\varkappa$ is regular if and only if, for every $\gamma<\varkappa$ and every family $(S_\alpha : \alpha<\gamma)$ of sets with $|S_\alpha| < \varkappa$ for any $\alpha<\gamma$, we have $|\bigcup\{S_\alpha : \alpha < \varkappa\}| < \varkappa$.

Proof. Let $\varkappa$ be singular, $\lambda = cf(\varkappa)$, and $f : \lambda \to \varkappa$ such that $\sup rng\, f = \varkappa$. Define $S_\alpha = f(\alpha)$. We have $\lambda<\varkappa$, $|S_\alpha| < \varkappa$ for every $\alpha<\lambda$, and $|\bigcup\{S_\alpha : \alpha<\lambda\}| = \varkappa$. Again by contraposition we prove the other direction. Let $(S_\alpha : \alpha<\gamma)$ be a family with $\gamma<\varkappa$, $|S_\alpha| < \varkappa$ for every $\alpha<\gamma$, and $|\bigcup\{S_\alpha : \alpha<\gamma\}| = \varkappa$. Let $\mu = \sup\{|S_\alpha| : \alpha<\gamma\}$. We show that $\mu = \varkappa$, and so we get, by $f(\alpha) = |S_\alpha|$, a function $f : \gamma \to \varkappa$ whose range is cofinal in $\varkappa$, which implies $cf(\varkappa) \leq \gamma < \varkappa$, and so $\varkappa$ is singular. If $\mu<\varkappa$, then $\varkappa = |\bigcup\{S_\alpha : \alpha<\gamma\}| < |\gamma| \cdot \mu$ by Lemma 5.10.(d), but $|\gamma| \cdot \mu = \max\{|\gamma|, \mu\} < \varkappa$, a contradiction.

Corollary 6.3. An infinite cardinal $\varkappa$ is regular if and only if it is not the sum of a family of less than $\varkappa$ cardinals which are smaller than $\varkappa$.

Definition. If $\varkappa$ is a cardinal, then we say that $C \subseteq \varkappa$ is closed in $\varkappa$ if, for any non-empty subset D of C, $\sup D \in C \cup \{\varkappa\}$. A closed and unbounded subset of $\varkappa$ is called a club in $\varkappa$.

If $(A,<)$ is a well ordering and $f : \alpha \to A$, we call f continuous if, for any limit ordinal $\beta<\alpha$, we have $f(\beta) = \sup\{f(\gamma) : \gamma<\beta\}$. f is called a normal function if f is strictly increasing and continuous.

Remark.

1) If $\varkappa$ is a regular cardinal, then $C \subseteq \varkappa$ is closed if and only if, for any sequence $(a_\alpha : \alpha<\lambda)$ of elements of C with length $\lambda<\varkappa$, we have $\sup\{a_\alpha : \alpha<\lambda\} \in C$.
2) If $\varkappa>\omega$ is regular, then the set $C_1 = \{\alpha<\varkappa : \alpha \text{ is a limit ordinal}\}$ is a club in $\varkappa$. Obviously C_1 is closed. Let $\alpha<\varkappa$, $\alpha \geq \omega$. Then $\varkappa > |\alpha| = |\alpha \cup \{\alpha+n : n \in \omega\}| = |\alpha+\omega|$, hence $\alpha+\omega<\varkappa$. Since $\alpha+\omega \in C_1$, C_1 is unbounded.
3) If $(A,<)$ is a well-ordering and $f : \alpha \to A$ is an isomorphism, then f is a normal function.

Lemma 6.4.
Let $\varkappa > \aleph_0$ be a regular cardinal.

(a) $C \subseteq \varkappa$ is a club in $\varkappa$ if and only if C is the range of a normal function $f : \varkappa \to \varkappa$.

(b) If $f : \varkappa \to \varkappa$ is a normal function, then the set of fixed points of f is a club in $\varkappa$.

Proof. For (a), let $C \subseteq \varkappa$ be a club in $\varkappa$, and let $f : \alpha \to C$ be the order isomorphism corresponding to the well ordering $(C, \in)$. For any $\gamma < \alpha$ we have $\gamma \leq f(\gamma)$, since f is strictly increasing (the smallest γ with $f(\gamma) < \gamma$ yields $f(f(\gamma)) < f(\gamma)$, a contradiction to the choice of γ). Therefore $\alpha \leq \varkappa$. Since $|C| = \varkappa$, we have $\alpha \geq \varkappa$. Hence $\alpha = \varkappa$. We show that $f : \varkappa \to \varkappa$ is continuous. Let $\beta < \varkappa$ be a limit ordinal. Since f is increasing, we have $\gamma = \sup\{f(\alpha) : \alpha < \beta\} \leq f(\beta)$. Further $\gamma \in C$ since C is closed. Choose $\sigma < \varkappa$ with $f(\sigma) = \gamma$. Since $f(\sigma) \leq f(\beta)$, we have $\sigma \leq \beta$. If $\sigma < \beta$, then $\sigma + 1 < \beta$ and $\gamma = f(\sigma) < f(\sigma+1) \leq \gamma$, a contradiction. So $\sigma = \beta$ and $\gamma = f(\beta)$. We have shown that $f : \varkappa \to \varkappa$ is a normal function with $\mathrm{rng}\, f = C$. On the other hand, if $f : \varkappa \to \varkappa$ is a normal function, then $\mathrm{rng}\, f$ is unbounded since $\gamma \leq f(\gamma)$ for every $\gamma < \varkappa$, and $\mathrm{rng}\, f$ is closed since f is continuous.
For (b), let $C = \{\alpha : f(\alpha) = \alpha\}$. Obviously C is closed since f is continuous. We show that C is unbounded. Let $\gamma < \varkappa$, $\alpha_0 = f(\gamma + 1)$, $\alpha_{n+1} = f(\alpha_n)$ for $n \in \omega$, and $\alpha = \sup\{\alpha_n : n \in \omega\}$. Then $\alpha > \gamma$, since f is strictly increasing, and, since f is continuous, $f(\alpha) = \sup\{f(\alpha_n) : n \in \omega\} = \sup\{\alpha_{n+1} : n \in \omega\} = \alpha$.

Lemma 6.5. If $\varkappa > \aleph_0$ is a regular cardinal and $(C_\alpha : \alpha < \gamma)$ a sequence of less than $\varkappa$ clubs in $\varkappa$, then $C = \bigcap\{C_\alpha : \alpha < \gamma\}$ is a club in $\varkappa$.

Proof. It is clear that C is closed. By transfinite induction on γ we show that C is unbounded. Let $\sigma < \varkappa$. If $\gamma = 2$, choose $\alpha_n \in C_0$ and $\beta_n \in C_1$ such that $\sigma < \alpha_0 < \beta_0 < \alpha_1 < \beta_1 < \ldots$. We have $\sigma < \alpha = \sup\{\alpha_n ; n \in \omega\} = \sup\{\beta_n : n \in \omega\} \in C_0 \cap C_1$, so $C_0 \cap C_1$ is unbounded. If $\gamma = \beta + 1$, the same proof works for $\bigcap\{C_\alpha : \alpha < \beta\}$ and C_β by the induction hypothesis. Now let γ be a limit ordinal. Choose $\eta_0 \in C_0$ with $\eta_0 > \sigma$ and $\eta_\nu \in \bigcap\{C_\alpha : \alpha < \nu\}$ such that $\eta_\nu > \sup\{\eta_\alpha : \alpha < \nu\}$ for every $\nu < \gamma$ - this is possible by the induction hypothesis and since $\varkappa$ is regular. Then we have $\eta := \sup\{\eta_\nu : \nu < \gamma\} > \sigma$. If $\nu_0 < \gamma$, then $\eta = \sup\{\eta_\nu ; \nu_0 < \nu < \gamma\}$ and $\eta_\nu \in C_{\nu_0}$ for any $\nu > \nu_0$, hence $\eta \in C_{\nu_0}$. So we have shown that $\eta \in \bigcap\{C_\alpha : \alpha < \gamma\} = C$, and C is unbounded.

Lemma 6.5 obviously is false if we admit sequences of length $\varkappa$: Let $C_\alpha = \varkappa \setminus \alpha$ for each $\alpha < \varkappa$. Then $\bigcap\{C_\alpha : \alpha < \varkappa\} = \emptyset$. But a new kind of intersection gives a positive result for $\varkappa$ clubs:

Lemma 6.6. Let $\varkappa > \aleph_0$ be a regular cardinal and $(C_\alpha : \alpha < \varkappa)$ a sequence of clubs in $\varkappa$. Let

$C = \Delta\{C_\alpha : \alpha < \varkappa\} = \{\beta < \varkappa : \beta \in C_\alpha \text{ for every } \alpha < \beta\}$ be the so-called diagonal intersection of $(C_\alpha : \alpha < \varkappa)$. Then C is a club in $\varkappa$.

Proof. First we show that C is closed. Let $(a_\alpha : \alpha < \gamma)$ be a sequence of elements of C with $\gamma < \varkappa$ and $\mathrm{Lim}(\gamma)$. Without loss of generality we can assume that $a_\alpha < a_\beta$ for $\alpha < \beta < \gamma$ - otherwise take an increasing subsequence with the same supremum. Let $\beta = \sup\{a_\alpha : \alpha < \gamma\}$. If $\sigma < \beta$, then $\beta = \sup\{a_\alpha : \alpha < \gamma \text{ and } \sigma < a_\alpha\}$, since β is a limit ordinal, and further $a_\alpha \in C_\sigma$ for $\sigma < a_\alpha$ and $\alpha < \gamma$, since $a_\alpha \in C$. Hence $\beta \in C_\sigma$, since C_σ is closed. We have shown that $\beta \in C_\sigma$ for every $\sigma < \beta$, and this means $\beta \in C$. To prove that C is unbounded, let $\sigma < \varkappa$. Choose $\alpha_0 \in C_0 \setminus \sigma$ and $\alpha_{n+1} \in \bigcap\{C_\alpha : \alpha < \alpha_n\} \setminus \alpha_n$ - this is possible by Lemma 6.5. The sequence $(\alpha_n : n \in \omega)$ is strictly increasing and its supremum β is smaller than $\varkappa$. If $\beta \in C$, we are done, since $\beta > \sigma$. Let $\nu < \beta$; we have to show that $\beta \in C_\nu$. Since β is a limit ordinal, we have $\beta = \sup\{\alpha_n : n > n_0\}$ for some n_0 such that $\alpha_n > \gamma$ for any $n \geq n_0$. But $\alpha_n \in C_\nu$ for every $n > n_0$, since $\nu < \alpha_{n_0}$ and by the definition of the sequence $(\alpha_n : n \in \omega)$. Hence $\beta \in C_\nu$, since C_ν is closed.

Definition. Let $\varkappa > \aleph_0$ be a regular cardinal. A subset S of $\varkappa$ is called stationary (in $\varkappa$), if $S \cap C \neq \emptyset$ for every club C in $\varkappa$. If $D \subseteq \varkappa$ and $f : D \to \varkappa$ has the property that $f(\alpha) < \alpha$ for every $\alpha \in D \setminus \{0\}$, we say that f is regressive on D.

Remark.

1) If C is a club in $\varkappa$ and if $S \subseteq \varkappa$ is stationary, then $C \cap S$ is stationary (Lemma 6.5).
2) If $S \subseteq \varkappa$ is stationary in $\varkappa$, then S is unbounded in $\varkappa$: S must meet every club $\varkappa \setminus \alpha$, $\alpha < \varkappa$.
3) The set $S = \{\alpha < \aleph_2 : \mathrm{cf}(\alpha) = \omega\}$ is stationary in $\aleph_2$, but not a club in $\aleph_2$: $\sup\{\beta < \aleph_1 : \mathrm{cf}(\beta) = \omega\} = \aleph_1$ is not an element of S by Lemma 6.1, so S is not closed.

We can reformulate Lemma 6.5 as follows:

Corollary 6.7.
Let $\kappa > \aleph_0$ be a regular cardinal.

(a) If $(D_\alpha : \alpha < \gamma)$ is a sequence of less than κ non-stationary subsets of κ, then $D = \bigcup\{D_\alpha : \alpha < \gamma\}$ is not stationary.

(b) If $S \subseteq \kappa$ is stationary, and $(S_\alpha : \alpha < \gamma)$ is a partition of S in less than κ classes (i.e. $S = \bigcup\{S_\alpha : \alpha < \gamma\}$ and $S_\alpha \cap S_\beta = \emptyset$ for $\alpha < \beta < \gamma$), then there exists an $\alpha_0 < \gamma$ such that S_{α_0} is stationary.

Theorem 6.8. (Fodor's Theorem)
If $\kappa > \aleph_0$ is a regular cardinal, $S \subseteq \kappa$ is stationary, and $f : S \to \kappa$ is regressive, then there exists a $\gamma < \kappa$ such that the set $\{\alpha \in S : f(\alpha) = \gamma\}$ is stationary in κ.

Proof. We assume that the proposition is false and choose κ, S, and f for which the theorem fails. Then we get, for each $\gamma < \kappa$, a club $C_\gamma \subseteq \kappa$ such that $f(\alpha) \neq \gamma$ for all $\alpha \in C_\gamma$. Let $C = \Delta\{C_\gamma : \gamma < \kappa\}$. Since C is a club by Lemma 6.6, $C \cap S$ is stationary. Choose $\alpha \in C \cap S \setminus \{0\}$ and let $\beta = f(\alpha)$. Then $\beta < \alpha$, since f is regressive, and $\alpha \notin C_\beta$ by the choice of C_β. But this contradicts $\alpha \in C$. Thus our theorem is proved.

Lemma 6.9. (Choice Lemma)
If $\mathcal{D} = (D_\alpha : \alpha < \kappa)$ is a family of non-stationary subsets of the regular cardinal $\kappa > \aleph_0$ and if $S = \bigcup\{D_\alpha : \alpha < \kappa\}$ is stationary in κ, then there are a subfamily $\mathcal{D}' = (D_{\alpha_\nu} : \nu < \kappa)$ of $\mathcal{D}$ and an injective choice function g for $\mathcal{D}'$ such that $\operatorname{rng} g$ is stationary in κ.

Proof. By transfinite induction, we define an injective sequence $(\alpha_\nu : \nu < \kappa)$, a strictly increasing sequence $(\beta_\nu : \nu < \kappa)$ with $\beta_\nu \in D_{\alpha_\nu}$, and a descending sequence $(C_\nu : \nu < \kappa)$ of clubs in κ such that $\beta_\nu \in \bigcap\{C_\alpha : \alpha < \nu\}$ and $C_\nu \cap \bigcup\{D_{\alpha_\gamma} : \gamma < \nu\} = \emptyset$ for any $\nu \in \kappa \setminus \{0\}$. The function g with $g(D_{\alpha_\nu}) = \beta_\nu$ will have the desired properties.

Let $\alpha_0 = \min\{\alpha : D_\alpha \neq \emptyset\}$, $\beta_0 \in D_{\alpha_0}$, and $C_0 = \kappa$. Now let α_γ, β_γ and D_γ be defined for every $\gamma < \nu$ and let first $\nu = \sigma + 1$ be a successor ordinal. $C_\sigma \cap S$ is stationary and, by the induction hypothesis, disjoint from $\bigcup\{D_{\alpha_\gamma} : \gamma < \sigma\}$. Further, by 6.7, $C_\sigma \cap S \setminus D_{\alpha_\sigma}$ is stationary.

Especially this set is unbounded, and therefore it has an element $\beta_\nu > \beta_\sigma$ such that $\beta_\nu \in D_{\alpha_\nu}$ for some $\alpha_\nu \notin \{\alpha_\gamma : \gamma \leq \sigma\}$. Again by 6.7, we find a club C^*_ν with $C^*_\nu \cap \bigcup\{D_{\alpha_\gamma} : \gamma \leq \nu\} = \emptyset$, and define $C_\nu = C^*_\nu \cap C_\sigma$.

Now let ν be a limit ordinal. Define $C_\nu = \bigcap\{C_\gamma : \gamma < \nu\}$ - C_ν is a club by Lemma 6.5. Take $\beta = \sup\{\beta_\gamma : \gamma < \nu\}$. Case 1: $\beta \subset C_\nu \cap S$. This is the crucial point in the construction, as we shall see in the second part of the proof. Note that $\beta \notin D_{\alpha_\gamma}$ for any $\gamma < \nu$ by the choice of $C_{\alpha_\gamma} \supseteq C_\nu$. Define $\beta_\nu = \beta$. Case 2: $\beta \notin C_\nu \cap S$. Then choose $\beta_\nu \in C_\nu \cap S$ as in the successor step with $\beta_\nu > \beta$. In both cases, we get $\beta_\nu \subset D_{\alpha_\nu}$ for some $\alpha_\nu \notin \{\alpha_\gamma : \gamma < \nu\}$. This completes the definition, and it is easy to see that our constructed objects have the desired properties.

If we define $g(D_{\alpha_\nu}) = \beta_\nu$, then g is an injective choice function for $\vartheta' = (D_{\alpha_\nu} : \nu < \varkappa)$ and $S_1 = \text{rng } g$ is unbounded in $\varkappa$. To show that S_1 is stationary, we define $\overline{S}_1$ as the set of all suprema of sequences in S_1 which are of length less than $\varkappa$. It is not difficult to see that $\overline{S}_1$ is closed, so $\overline{S}_1$ is a club and $\overline{S}_1 \cap S$ is stationary in $\varkappa$. Consequently it suffices to prove the following

Claim. $S_1 = \overline{S}_1 \cap S$.

By the definition of $\overline{S}_1$, we have $S_1 \subseteq \overline{S}_1 \cap S$. For the proof of the converse inclucion let $\gamma \in \overline{S}_1 \cap S$, $\gamma = \sup\{\beta_{\nu_\sigma} : \sigma < \rho\}$, $\beta_{\nu_\sigma} \in S_1$ for any $\sigma < \rho$, $\rho < \varkappa$. Without loss of generality we can assume that $(\beta_{\nu_\sigma} : \sigma < \rho)$ is strictly increasing and ρ is a limit ordinal. If we take $\delta = \sup\{\nu_\sigma : \sigma < \rho\}$, then $\gamma = \sup\{\beta_\sigma : \sigma < \delta\}$. We want to show that $\gamma \in C_\delta$, because this implies that $\gamma = \beta_\delta$ by the construction of the sequence $(\beta_\nu : \nu < \varkappa)$, since $\gamma \in S$ and δ is a limit ordinal. For this reason we show $\gamma \in C_\alpha$ for any $\alpha < \delta$. Let $\alpha < \delta$. Then there exists a $\sigma_0 < \rho$ such that $\alpha < \nu_\sigma$ for each $\sigma \geq \sigma_0$, and consequently, by our definition, $\beta_{\nu_\sigma} \in C_\alpha$ for each $\sigma \geq \sigma_0$. Therefore $\gamma = \sup\{\beta_{\nu_\sigma} : \sigma_0 \leq \sigma < \rho\} \in C_\alpha$, since C_α is closed. So we have $\gamma \in \bigcap\{C_\alpha : \alpha < \delta\} = C_\delta$, and the proof is completed.

Chapter II

GENERAL CRITERIA AND THEIR APPLICATIONS

§1. *Some basic definitions of graph theory*

If A and B are sets, then we define $A \otimes B = \{\{a,b\} : a \in A, b \in B,\ a \neq b\}$ to be the *set of unordered pairs* of elements of A and B. An ordered pair (V,E) of sets such that $E \subseteq V \otimes V$ is called a *graph*. V is the *set of vertices* and E is the *set of edges* of the graph $G = (V,E)$. A vertex $v \in V$ is said to be *incident* with an edge $e \in E$ if $v \in e$. $d(v) = |\{x \in V : \{v,x\} \in E\}|$ is the *degree* of the vertex v. The vertex v is an *endvertex* if $d(v) = 1$. Denote by $V(G)$ the vertex set of the graph G and by $E(G)$ the edge set of G. A graph G_1 is a *subgraph* of a graph G_2 if $V(G_1) \subseteq V(G_2)$ and $E(G_1) \subseteq E(G_2)$. If G_1 is a subgraph of G_2, we write $G_1 \leq G_2$. If $G = (V,E)$ is a graph and A is a subset of V, then $G[A] = (A,(A \otimes A) \cap E)$ is the subgraph of G *induced* by A. If $S \subseteq V$, then $G \setminus S = G[V \setminus S]$ is the graph which one gets by deleting all vertices of S and all edges which are incident with a vertex of S. Let $k \in \omega \cup \{\omega, \mathbb{Z}\}$. An injective sequence $p = (v_i : i \in k)$ of vertices of a graph is called a *path*, if $\{\{v_i, v_{i+1}\} : i+1 \in k\} \subseteq E$. If $k \in \omega$, $k = \omega$ or $k = \mathbb{Z}$, then the path p is called *finite*, *one-way infinite* and *two-way infinite* respectively. If $p = (v_i : i \in k)$ is a path, then $V(p) = \{v_i : i \in k\}$ is the set of vertices of p and $E(p) = \{\{v_i, v_{i+1}\} : i+1 \in k\}$ is the set of edges of p. Let $p = (v_i : i < k)$ be a finite path. Then v_0 is called the *starting point* and v_{k-1} the *endpoint* of p. We define on the vertex set V of a graph G an equivalence relation $\sim$ as follows: $x \sim y$ if and only if there is a path $(v_i : i \leq k)$ such that $x = v_0$ and $y = v_k$. If $A \subseteq V$ is an equivalence class of $\sim$, the graph $G[A]$ is called a *connected component* of G or simply a component of G. Every graph is the disjoint union of its components. A graph which has exactly one component is said to be a *connected graph*. If $f \subseteq E$ and $x \in V$, then $f\langle x\rangle = \{y \in V : \{x,y\} \in f\}$. If $|f\langle x\rangle| = 1$, then $f(x)$ denotes the unique element of $f\langle x\rangle$. For $X \subseteq V$ let $f[X] = \bigcup\{f\langle x\rangle : x \in X\}$ and $f \restriction X = f \cap (X \otimes V)$. A set $f \subseteq E$ is called a *matching* in G if the edges of f are pairwise disjoint. A matching f is said to be a matching of a subset A of V if every edge of f is incident with an element of A and if, for every element $a \in A$, there is an edge of f incident with a. A set $A \subseteq V$ is matchable if there exists a matching of A. If $B \subseteq V$, then A is called matchable into B if there exists a matching f of A such that $f[A] \subseteq B$. Let f be a matching in G. We call a path $p = (v_i : i \in k)$ f-alternating, if the edges $\{v_i, v_{i+1}\}$ are alternately in f and in $E \setminus f$.

A bipartite graph G is a graph whose vertex set $V(G)$ can be partitioned into two subsets M and W such that every edge of G is an element of $M \otimes W$. We denote a bipartite graph by $\Gamma = (M,W,E)$. To the elements of M and W respectively we refer as men and women, and we say that $m \in M$ knows $w \in W$ if $\{m,w\} \in E$. If $\Gamma = (M,W,E)$ is a bipartite graph, $J \subseteq M$ and $A \subseteq W$, then $\Gamma(J,A)$ denotes the induced subgraph $(J,A,(J \otimes A) \cap E)$. In the bipartite case, "subgraph" always means "induced subgraph". We write $\Pi \leq \Gamma$ if $\Pi = \Gamma(J,A)$ for some $J \subseteq M$ and $A \subseteq W$. We say that the subgraph $\Pi = \Gamma(J,A)$ of Γ is saturated in Γ if $E_\Gamma(J) \subseteq A$, i.e. every man in Π doesn't know a woman in Γ whom he not already knows in Π. The graph $\tilde{\Gamma} = (W,M,E)$ is called the dual graph of $\Gamma = (M,W,E)$. A marriage of a bipartite graph $\Gamma = (M,W,E)$ is a matching of M (in Γ). For $X \subseteq W$ we call $D_\Gamma(X) = \{m \in M : E<m> \subseteq X\}$ the demand of X. If f is a matching of $D_\Gamma(X)$, then every $m \in D_\Gamma(X)$ is married with an element of X.

Let $F = (F(i) : i \in I)$ be a family of sets with index set I such that $I \cap \bigcup\{F(i) : i \in I\} = \emptyset$. If $E = \{\{i,x\} : i \in I, x \in F(i)\}$, then the graph $\Gamma_F = (I, \bigcup\{F(i) : i \in I\}, E)$ is called the bipartite graph associated with F [1]. Conversely let $\Gamma = (M,W,E)$ be a bipartite graph and $F(m) = \{w \in W : \{m,w\} \in E\}$ for every $m \in M$. The family $F_\Gamma = (F(m) : m \in M)$ is the family associated with Γ. Because of this natural correspondence between bipartite graphs and families we often do not distinguish between these two notions. So an injective choice function for a family F is regarded as a marriage of Γ_F and vice versa. If we use choice functions for families in bipartite graphs, then we regard their elements as unordered pairs. In general we will define many notions only for families and will use them with their obvious meanings for bipartite graphs without further comments.

Throughout the following chapters, J,K,L always denote sets of men, and A,B,C,X,Y denote sets of women. m,i,j denote men, and w,a,b denote women.

---

[1] In all examples, the sets I and $F(I) = \bigcup\{F(i) : i \in I\}$ will not be disjoint. It is clear how this can be arranged.

If V is a set and $E \subseteq V \times V$ is an irreflexive, antisymmetric relation, then we call the ordered pair (V,E) a <u>directed graph</u>. An injective sequence $(v_i : i<k\leq\omega)$ of vertices of a directed graph (V,E) is said to be a <u>(directed)</u> <u>path</u>, if $(v_{i-1},v_i)\in E$ for all $i<k$. Finite and infinite directed paths are defined analogously to the undirected case.

§2. Critical sets and families

In this section we prove the basic facts on critical sets (see [St1],[PS1]). They are applied in every section of this book, and so we recommend to the reader to study them thoroughly.

Let us begin with some basic definitions. If $F = (F(i) : i\in I)$ is a family and f is an injective choice function for F, i.e. an injective function $f : I \to \bigcup\{F(i) : i\in I\}$ such that $f(i)\in F(i)$ for any $i\in I$, then we call f a <u>marriage</u> of F. We denote by $\mathcal{M}$ the class of all families F for which there exists a marriage of F. A function g is said to be a <u>matching</u> in the family $F = (F(i) : i\in I)$, if there is a subset $J\subseteq I$ such that g is a marriage of the subfamily $F \upharpoonright J$. A subset J of the domain of a family F is called <u>matchable</u> if $F\upharpoonright J\in\mathcal{M}$. If A, K, X, Y are sets, $F = (F(i) : i\in I)$ is a family, and f is a matching in F, then we use the following abbreviations:

$(F\setminus K)\setminus A = (F(i)\setminus A : i\in I\setminus K)$,
$F\setminus A = (F\setminus\emptyset)\setminus A$, if $A\cap I = \emptyset$,
$F\setminus f = (F\setminus \mathrm{dom}\, f)\setminus \mathrm{rng}\, f$,
$F^Y = (F(i) : F(i)\subseteq Y)$, and
$F_X = (F(i)\setminus X : F(i)\not\subseteq X)$.

If $X\subseteq Y$, then $F_X^Y = (F^Y)_X = (F_X)^Y$. It is possible that the family $F\setminus A$ has empty members, but for the family F_A this is impossible. Therefore the inclusion $F_A\subseteq F\setminus A$ often is a proper inclusion.

Furthermore let
$F(K) = \bigcup\{F(k) : k\in K\}$, and
$F^{-1}(A) = \{i\in I : F(i)\cap A \neq \emptyset\}$.
If $A = \{a\}$, then we write $F^{-1}(a)$ for $F^{-1}(\{a\})$.

We shall see that the following notion of a critical set is of basic importance. Let $F = (F(i) : i\in I)$ be a family. A subset K of I is called <u>critical in F</u>, if $F\upharpoonright K\in\mathcal{M}$ and if moreover $\mathrm{rng}\, f = f[K]$ for any marriage f of $F\upharpoonright K$. A subfamily $F\upharpoonright K$ is said to be <u>critical</u> if K is critical in F.

Let us now prove some basic facts about critical sets.

Lemma 2.1. If $F \in \mathcal{M}$ and if each element of $\mathcal{K}$ is critical in F, then $\cup\mathcal{K}$ and $\cap\mathcal{K}$ are critical in F.

Proof. Since F has a marriage, $F \upharpoonright \cup\mathcal{K}$ has a marriage. If f is an arbitrary marriage of $F \upharpoonright \cup\mathcal{K}$, then $f[K] = F(K)$ for all $K \in \mathcal{K}$. Therefore $f[\cup\mathcal{K}] = \cup\{f[K] : K \in \mathcal{K}\} = \cup\{F(K) : K \in \mathcal{K}\} = F(\cup\mathcal{K})$. Hence $\cup\mathcal{K}$ is a critical set in F.

Now we want to show that $\cap\mathcal{K}$ is critical in F. First we prove $f[\cap\mathcal{K}] = F(\cap\mathcal{K})$ for each marriage f of $F \upharpoonright \cup\mathcal{K}$. So let f be a marriage of $F \upharpoonright \cup\mathcal{K}$ and x be an element of $F(\cap\mathcal{K})$. Since $\cup\mathcal{K}$ is critical, there exists an $j \in \cup\mathcal{K}$ such that $f(j) = x$. Assume that $j \notin \cap\mathcal{K}$. Then there is a $K \in \mathcal{K}$ such that $j \notin K$. Since $x \in F(\cap\mathcal{K}) \subseteq F(K)$, $f \upharpoonright K$ is a marriage of $F \upharpoonright K$ with $\operatorname{rng} f \upharpoonright K \neq F(K)$ which contradicts the fact that K is critical in F. This shows that $j \in K$ and $x \in f[\cap\mathcal{K}]$. Now choose a marriage f of $F \upharpoonright \cup\mathcal{K}$ and consider an arbitrary marriage g of $F \upharpoonright \cap\mathcal{K}$. Since $f[\cap\mathcal{K}] = F(\cap\mathcal{K})$, the mapping $h = f \upharpoonright (\cup\mathcal{K} \setminus \cap\mathcal{K}) \cup g$ is a marriage of $F \upharpoonright \cup\mathcal{K}$. So $h[\cap\mathcal{K}] = g[\cap\mathcal{K}] = F(\cap\mathcal{K})$.

Definition.
Let K be critical in the family F.

1. K is called maximal critical in F if $K = L$ for any set $L \supseteq K$ which is critical in F.
2. K is said to be the greatest critical set of F if $L \subseteq K$ for all sets L which are critical in F.

Corollary 2.2. If a family F has a marriage, then there exists a greatest critical set in F.

Proof. Let $\mathcal{K}$ be the set of all critical sets in F. By Lemma 2.1, $\cup\mathcal{K}$ is critical, and so it is the greatest critical set in F.

If a family F has no marriage, then F will contain a maximal critical subfamily.

Lemma 2.3. If $(K_\alpha : \alpha<\gamma)$ is a chain of sets which are critical in the family F, then $\cup\{K_\alpha : \alpha<\gamma\}$ is critical in F.

Proof. Let $K = \cup\{K_\alpha : \alpha<\gamma\}$. By Lemma 2.1, it is enough to prove that $F \restriction K$ has a marriage. For $\alpha<\gamma$ let g_α be a marriage of $F \restriction K_\alpha$ and put $L_\alpha = K_\alpha \setminus \cup\{K_\beta : \beta<\alpha\}$. Since $g_\beta[K_\beta] = F(K_\beta) = g_\alpha[K_\beta]$ for all $\beta<\alpha$, we conclude $g_\beta[L_\beta] \cap g_\alpha[L_\alpha] = \emptyset$. Therefore $\cup\{g_\alpha \restriction L_\alpha : \alpha<\gamma\}$ is a marriage of $F \restriction K$.

Theorem 2.4. For each set K which is critical in the family F, there exists a maximal critical set L in F containing K. Therefore every family has a maximal critical subfamily.

Proof. The first assertion is proved by Zorn's Lemma. Since $\emptyset$ is critical in every family, the second assertion follows directly from the first one.

Lemma 2.5. If K is critical in $F \setminus A$ (in F_A) and if $B \supseteq A$, then there exists a subset L of K which is critical in $F \setminus B$ (in F_B) such that $F(L) \cup B = F(K) \cup B$.

Proof. If f is a marriage of $(F \restriction K) \setminus A$, then put $L = \{i \in K : f(i) \notin B\}$.

Lemma 2.6. If K is critical in F and L is critical in $F \setminus F(K)$, then $K \cup L$ is critical in F.

Proof. Let f and g be marriages of the families F and $(F \restriction L) \setminus F(K)$ respectively. Then $f \cup g$ is a marriage of $F \restriction K \cup L$. If h is an arbitrary marriage of $F \restriction K \cup L$, then $h[K] = F(K)$, and therefore $h \restriction L$ is a marriage of $(F \restriction L) \setminus F(K)$. It follows that $h[L] = F(L) \setminus F(K)$. This proves $h[K \cup L] = F(K \cup L)$.

Theorem 2.7. If N is a maximal critical set in a family F and K is critical in F, then $F(K) \subseteq F(N)$.

Proof. Let f be a marriage of $F \restriction K$ and $L = \{i \in K : f(i) \notin F(N)\}$. L is critical in $F \setminus F(N)$ and, by Lemma 2.6, $N \cup L$ is critical in F. Since N is maximal critical, it follows that L is empty. Therefore $f[K] = F(K) \subseteq F(N)$.

Lemma 2.8. If K is a critical set in a family F and if f is a marriage of $F \restriction K$, then the bipartite graph Γ_F associated with F has no infinite f-alternating path.

Proof. Let us assume that Γ_F possesses a one-way infinite f-alternating path p. Without loss of generality let the first vertex of p be an element of $F(K)$ and the first edge of p be an element of f. Then $f \Delta E(p) = (f \setminus E(p)) \cup (E(p) \setminus f)$ is a marriage of $F \restriction K$ which leaves the first vertex of p unmarried, contradicting the fact that K is critical in F.

Corollary 2.9. Every nonempty critical family has a finite member.

Proof. Let F be a nonempty critical family and let us assume that $|F(i)| \geq \aleph_0$ for all $i \in \operatorname{dom} F = K$. Take a marriage f of F and define by recursion an infinite sequence $(i_n : n \in \omega)$ of elements of K. Choose $i_0 \in K$ and let $i_{n+1} \in K$ such that $f(i_{n+1}) \in F(i_n) \setminus \{f(i_k) : k \leq n\}$. The sequence $(i_0, f(i_1), i_1, f(i_2), \ldots)$ is an infinite f-alternating path and, by Lemma 2.8, this contradicts the assumption that F is a critical family.

Remark: An analogous proof shows that every infinite critical family has infinitely many finite members. Furthermore we demonstrate in Section III.7 that critical families are behaving essentially as those critical families which possess finite members only.

Theorem 2.10. Let K be critical in a familiy $F \setminus A$, and let $F(J) \cup B \subseteq F(K) \cup A$. If $(F \restriction J) \setminus B$ has a marriage, then

$$|B \setminus A| + |J \setminus K| \leq |A \setminus B| + |K \setminus J|.$$

Therefore

$$|B| + |J \setminus K| \leq |A| + |K \setminus J| \text{ and}$$

$$|B| + |J| \leq |A| + |K|.$$

Proof. Let f and g be marriages of $(F \upharpoonright K)\setminus A$ and $(F \upharpoonright J)\setminus B$ respectively. For each $x \in (B\setminus A)\cup(J\setminus K)$, we define by recursion a sequence $(x_n : n<\omega)$. If $x \in B\setminus A$, then put $x_0 = x$, $x_1 = f^{-1}(x_0)$, and if $x \in J\setminus K$, let $x_0 = x$, $x_1 = g(x_0)$. Put

$$x_{n+1} = \begin{cases} g(x_n) & \text{if } x_n = f^{-1}(x_{n-1}) \text{ and } x_n \in J\cap K, \\ f^{-1}(x_n) & \text{if } x_n = g(x_{n-1}) \text{ and } x_n \in F(K)\setminus A, \\ x_n & \text{otherwise} \end{cases}.$$

We hope that the following diagram will clarify the situation.

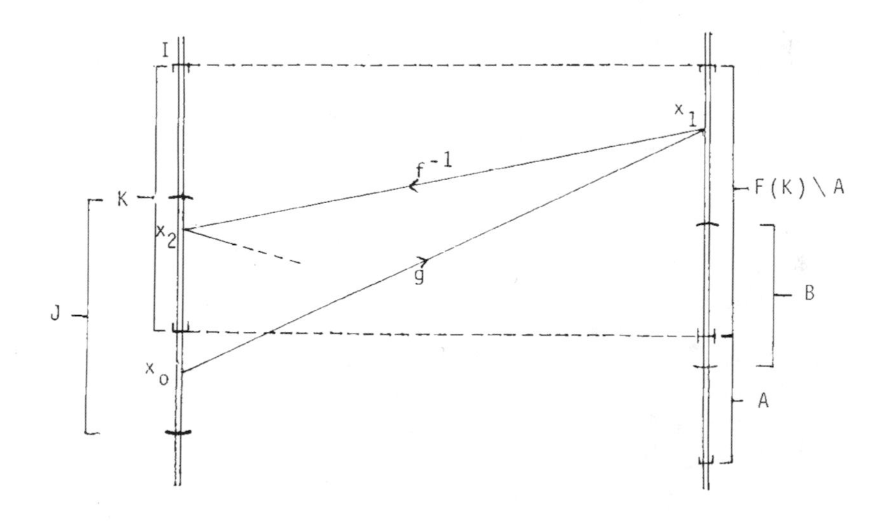

For each such sequence $(x_n : n<\omega)$, starting with x, there exists, by Lemma 2.8, a least natural number n, denoted by k_x, such that $x_n = x_{n+1}$. In addition, if $x,y \in (B\setminus A)\cup(J\setminus K)$ are different and $(x_n : n\le k_x)$, $(y_n : n\le k_y)$ are the associated sequences respectively, then $\{x_n : n\le k_x\}\cap\{y_n : n\le k_y\} = \emptyset$, since f and g are injective functions. Therefore the function

$h : (B\setminus A)\cup(J\setminus K) \to (A\setminus B)\cup(K\setminus J)$,

defined by $h(x) = x_{k_x}$, is injective.

Theorem 2.11. (Main Lemma)

If the family F has a marriage and if the family $F\setminus\{a\}$ has no marriage, then there exists a set K critical in F such that $a \in F(K)$.

Proof. Let f be a marriage of F. We define by recursion two sequences $(S_n : n\in\omega)$ and $(J_n : n\in\omega)$ such that $S_n\subseteq F(I)$ and $J_n\subseteq I$ for all $n\in\omega$. Put

$$S_0 = \{a\},\ J_n = f^{-1}[S_n],\ S_{n+1} = F(J_n)$$

for any $n\in\omega$. Now define

$$K = \cup\{J_n : n\in\omega\}.$$

Claim 1. If g is a marriage of F, then $S_n\subseteq \operatorname{rng} g$ for all $n\in\omega$.

Let us assume that there is a natural number n and a marriage g of F such that $S_n\setminus\operatorname{rng} g \neq \emptyset$. Let n be minimal with this property. By the hypothesis of Theorem 2.11, $n>0$. So if $x\in S_n\setminus\operatorname{rng} g$, then there is an $i\in J_{n-1}$ such that $x\in F(i)$. Put $k = n-1$. Of course $f(i)\in S_k$. We distinguish two cases.

Case 1. $f(i) = g(i)$.

Then $h := (g\setminus\{(i,g(i))\})\cup\{(i,x)\}$ is a marriage of F such that $f(i)\notin\operatorname{rng} h$. Therefore $S_k\not\subseteq\operatorname{rng} h$, contradicting the minimality of n.

Case 2. $f(i) \neq g(i)$.

Define recursively a sequence $(i_m : m<l\leq\omega)$ of elements of I by $i_0 = i$ and $i_{m+1} = f^{-1}(g(i_m))$, if $g(i_m)\in\operatorname{rng} f$. If $i_0 = i_{m+1}$ for some $m<\omega$, then $g' := (g\setminus\{(i_m,g(i_m)) : m<\omega\})\cup\{(i_m,f(i_m)) : m<\omega\}$ is a marriage of F such that $x\notin\operatorname{rng} g'$ and $f(i_0) = g'(i_0)$. We have just proved, see Case 1, that this leads to a contradiction. Otherwise $i_0 \neq i_{m+1}$ for each natural number m with $m+1<l$. The function $h = (f\setminus\{(i_m,f(i_m)) : m<l\})\cup\{(i_m,g(i_m)) : m<l\}$ is a marriage of F such that $f(i)\notin\operatorname{rng} h$. Therefore $S_k\not\subseteq\operatorname{rng} h$, contradicting the minimality of n.

Claim 2. K is critical in F.

Let us assume that there is a marriage h of $F\restriction K$ such that $\operatorname{rng} h\subsetneq F(K)$. Then there is a natural number n such that $S_n\not\subseteq\operatorname{rng} h$. By the construction of K, $\operatorname{rng}(f\restriction(I\setminus K))\cap F(K) = \emptyset$. Therefore $g = h\cup f\restriction(I\setminus K)$ is a marriage of F such that $S_n\not\subseteq\operatorname{rng} g$, contradicting Claim 1.

Of course $a\in F(K)$ by the construction of K, and the theorem is proved.

Corollary 2.12. If the family F has a marriage and K is maximal critical in F, then $F\setminus\{a\}$ has a marriage for any $a\in F(I)\setminus F(K)$.

Corollary 2.13. If $F \in \mathcal{M}$ and $A \subseteq \operatorname{rng} f$ for each marriage f of F, then there exists a set K critical in F such that $A \subseteq F(K)$.

Proof. Theorem 2.11 and Lemma 2.1.

§3. μ-tests

If F is a family and A is a set, then $\overline{A} = \cup\{A \cup F(K) : K$ critical in $F \setminus A\}$ is called the closure of A.

Lemma 3.1. If K is maximal critical in a family $F \setminus A$, then $\overline{A} = A \cup F(K)$.

Proof. If L is critical in $F \setminus A$, then $F(L) \setminus A \subseteq F(K) \setminus A$ by Theorem 2.7, and so $\overline{A} \subseteq A \cup F(K)$. The converse inclusion is trivial.

Lemma 3.2.
For any sets A and B,
(1) $A \subseteq \overline{A}$.
(2) $\overline{A} \subseteq \overline{B}$ if $A \subseteq B$.
(3) $\overline{\overline{A}} = \overline{A}$.

Proof. Let K be maximal critical in $F \setminus A$. By Lemma 3.1, $\overline{A} = A \cup F(K)$. (1) is now obvious. To prove (2), let A be a subset of B. By Lemma 2.5, there exists a set $L \subseteq K$ critical in $F \setminus B$ such that $F(K) \cup B = F(L) \cup B$. Therefore $\overline{A} \subseteq \overline{B}$. For (3), let $\overline{\overline{A}} = \overline{A} \cup F(K^*)$ where K^* is maximal critical in $F \setminus \overline{A}$. By Lemma 2.6, $K \cup K^*$ is critical in $F \setminus A$. Therefore $K^* = \emptyset$ by the maximality of K. This proves $\overline{\overline{A}} = \overline{A}$.

Lemma 3.3. If $\overline{X} \subseteq Y$, $a \in Y \setminus \overline{X}$, and $F_X^Y \in \mathcal{M}$, then F_X^Y has a marriage f such that $a \notin \operatorname{rng} f$.

Proof. Let K be maximal critical in F_X. Since $\overline{X} \subseteq Y$ and $F(K) \subseteq \overline{X}$, the set K is maximal critical in F_X^Y. Therefore $\overline{X} = X \cup F_X^Y(K)$. It follows that $a \notin F_X^Y(K)$. By Corollary 2.12, F_X^Y has a marriage f such that $a \notin \operatorname{rng} f$.

Lemma 3.4. If $\overline{X}\subseteq Y$ and $F_X^Y \in \mathcal{M}$, then $F_{\overline{X}}^Y \in \mathcal{M}$.

Proof. Let K be critical in $F\setminus X$ such that $\overline{X} = X\cup F(K)$. Since $\overline{X}\subseteq Y$, the set K is critical in F_X^Y. Let f be a marriage of F_X^Y. Then $f[K] = \overline{X}\setminus X$, and therefore $f\upharpoonright(\operatorname{dom} f\setminus K)$ is a marriage of $F_{\overline{X}}^Y$.

The notion of the closure of a set A is one of the basic ideas in Podewski's paper [P]. A further fundamental tool is the notion of a μ-test.

Definition.
A sequence $(X_\alpha : \alpha < \varkappa)$ of sets is called a continuous chain if
(1) $X_0 = \emptyset$,
(2) $X_\alpha \subseteq X_\beta$ if $\alpha < \beta < \varkappa$, and
(3) $X_\lambda = \cup\{X_\alpha : \alpha < \lambda\}$ for each limit ordinal λ.

Definition. Let F be a family and $\mu > \aleph_0$ a regular cardinal number. A continuous chain $(X_\alpha : \alpha < \mu)$ is called a μ-test of the family F if, for each ordinal $\alpha < \mu$, there exist a set A_α and a set K_α critical in $F\setminus A_\alpha$ such that $|A_\alpha| < \mu$ and $X_\alpha = A_\alpha \cup F(K_\alpha)$.

We need μ-tests for testing whether a family has a marriage or not. But first let us consider an important example of a μ-test.

Lemma 3.5. Let $\mu > \aleph_0$ be a regular cardinal number and $(A_\alpha : \alpha < \mu)$ a continuous chain such that $|A_\alpha| < \mu$ for all $\alpha < \mu$. Then the sequence $(X_\alpha : \alpha < \mu)$, defined by

$$X_\alpha = \cup\{\overline{A_{\beta+1}} : \beta < \alpha\},$$

is a μ-test which has the following properties: For each $\alpha < \mu$,
(1) $X_\alpha \subseteq \overline{A}_\alpha$.
(2) $\overline{X}_\alpha = \overline{A}_\alpha$.
(3) $\overline{X}_\alpha \subseteq X_{\alpha+1}$.

Proof. By definition, $(X_\alpha : \alpha < \mu)$ is a continuous chain.

Claim 1. $X_{\alpha+1} = \overline{A_{\alpha+1}}$ for each $\alpha < \mu$.

If $\beta<\alpha+1$, then $A_{\beta+1}\subseteq A_{\alpha+1}$ and so, by Lemma 3.2, $\overline{A}_{\beta+1}\subseteq\overline{A}_{\alpha+1}$. Therefore $X_{\alpha+1} = \overline{A}_{\alpha+1}$.

<u>Claim 2</u>. For any $\alpha<\mu$, there is a set K_α critical in $F\setminus A_\alpha$ such that $X_\alpha = A_\alpha\cup F(K_\alpha)$.

If $\alpha = 0$, then choose $K_\alpha = \emptyset$. If $\alpha = \beta+1$, then the claim follows from Claim 1 and Lemma 3.1. So let α be a limit ordinal and $\beta<\alpha$. Then there exists, by Lemma 3.1, a set N_β critical in $F\setminus A_{\beta+1}$ such that $\overline{A}_{\beta+1} = A_{\beta+1}\cup F(N_\beta)$. Since $A_{\beta+1}\subseteq A_\alpha$, there exists, by Lemma 2.5, a set $L_\beta\subseteq N_\beta$ critical in $F\setminus A_\alpha$ with $F(L_\beta)\cup A_\alpha = F(N_\beta)\cup A_\alpha$. Therefore

$$\begin{aligned} X_\alpha &= \bigcup\{\overline{A}_{\beta+1} : \beta<\alpha\} \\ &= \bigcup\{A_{\beta+1}\cup F(N_\beta) : \beta<\alpha\} \\ &= A_\alpha\cup\bigcup\{F(N_\beta) : \beta<\alpha\} \\ &= A_\alpha\cup\bigcup\{F(L_\beta) : \beta<\alpha\} \\ &= A_\alpha\cup F(\bigcup\{L_\beta : \beta<\alpha\}) . \end{aligned}$$

Now choose K_α maximal critical in $(F\restriction\bigcup\{L_\beta:\beta<\alpha\})\setminus A_\alpha$. Then $X_\alpha = A_\alpha\cup F(K_\alpha)$, which proves Claim 2.

By Claim 2 and the definition of $\overline{A}_\alpha$, we conclude $X_\alpha\subseteq\overline{A}_\alpha$. We get, by Lemma 3.2, $\overline{X}_\alpha\subseteq\overline{\overline{A}}_\alpha = \overline{A}_\alpha$. $A_\alpha\subseteq X_\alpha$ yields, again by Lemma 3.2, $\overline{A}_\alpha\subseteq\overline{X}_\alpha$, and so it follows that $\overline{X}_\alpha = \overline{A}_\alpha$. Since $A_\alpha\subseteq A_{\alpha+1}$, we have $\overline{X}_\alpha = \overline{A}_\alpha\subseteq\overline{A}_{\alpha+1} = X_{\alpha+1}$.

<u>Lemma 3.6</u>. Let $(X_\alpha:\alpha<\mu)$ be a μ-test and f a marriage of the family $F\restriction\{i\in I:\exists\alpha<\mu(F(i)\subseteq X_\alpha)\}$. Then there exists a club C in μ such that, for each $\alpha\in C$, the following holds: If $f(i)\in X_\alpha$, then there is an ordinal $\beta<\alpha$ such that $F(i)\subseteq X_\beta$. Clearly we can assume that $0\in C$.

<u>Proof</u>. Let us assume that Lemma 3.6 is false. Then

$$S = \{\alpha<\mu : \mathrm{Lim}(\alpha)\wedge\exists i(f(i)\in X_\alpha\wedge\forall\beta<\alpha\; F(i)\not\subseteq X_\beta)\}$$

is a stationary subset of μ. Since $(X_\alpha:\alpha<\mu)$ is continuous, there are, for every $\alpha\in S$, an ordinal $\beta_\alpha<\alpha$ and an $i_\alpha\in\mathrm{dom}\, f$ such that $f(i_\alpha)\in X_{\beta_\alpha}$ and $F(i_\alpha)\not\subseteq X_\sigma$ for each $\sigma<\alpha$. By Fodor's Theorem, there is an ordinal $\beta<\mu$ such that $\{\alpha\in S:\beta_\alpha = \beta\}$ is a stationary subset of μ. Therefore the set

$$\{\alpha\in S : f(i_\alpha)\in X_\beta\wedge F(i_\alpha)\not\subseteq X_\beta\}$$

is stationary in μ. We shall show that the set

$B = \{f(i) : f(i) \in X_\beta \wedge F(i) \not\subseteq X_\beta\}$
has cardinality μ. To prove this let $i \in \mathrm{dom}\, f$. Then there is an ordinal γ such that $F(i) \subseteq X_\gamma$. But, for each $\alpha \in S$ with $\gamma < \alpha$, we have $F(i_\alpha) \not\subseteq X_\gamma$. Hence $|\{\alpha \in S : i_\alpha = i\}| \leq |\gamma| < \mu$. Since μ is a regular cardinal, we conclude that $|B| = \mu$.

By the definition of a μ-test, there exist a set A of cardinality less than μ and a set K critical in $F \setminus A$ such that $X_\beta = A \cup F(K)$. Since $F(K) \subseteq X_\beta$, K is a subset of the domain of the marriage f and $f \restriction K$ is a marriage of $(F \restriction K) \setminus B$. Since $B \subseteq X_\beta = A \cup F(K)$, we conclude $F(K) \cup B \subseteq F(K) \cup A$. Now Theorem 2.10 yields $|B| \leq |A| < \mu$. This contradicts the fact that $|B| = \mu$, and the lemma is proved.

<u>Corollary 3.7</u>. Suppose that the family F has a marriage and $(X_\alpha : \alpha < \mu)$ is a μ-test of F. Then there exists a club C in μ with $0 \in C$ such that, for each $\alpha \in C$ and each member $F(i) \subseteq X_\alpha$, there is an ordinal $\beta < \alpha$ such that $F(i) \subseteq X_\beta$.

<u>Corollary 3.8</u>. Suppose that the family F has a marriage and $(X_\alpha : \alpha < \mu)$ is a μ-test of F. Then there is a club C in μ so that $0 \in C$ and

$$F_{X_\alpha}^{X_\beta} \text{ has a marriage}$$

for each $\alpha \in C$ and for each ordinal $\beta \geq \alpha$.

<u>Proof</u>. Let f be a marriage of the family $F \restriction \{i \in I : \exists \alpha < \mu\ F(i) \subseteq X_\alpha\}$. By Lemma 3.6, there is a club C such that $0 \in C$ and $F(i) \subseteq X_\alpha$, if $\alpha \in C$ and $f(i) \in X_\alpha$. Now let $\alpha \in C$ and $\beta \geq \alpha$. Then $f \restriction \{i \in I : F(i) \not\subseteq X_\alpha \text{ and } F(i) \subseteq X_\beta\}$ is a marriage of $F_{X_\alpha}^{X_\beta}$.

<u>Definition</u>.
A μ-test $(X_\alpha : \alpha < \mu)$ of a family F is called <u>positive</u> if there exists a club C in μ with $0 \in C$ such that, for each $\alpha \in C$, the following holds:

(1) If $F(i) \subseteq X_\alpha$, then there is an ordinal $\beta < \alpha$ such that $F(i) \subseteq X_\beta$.

(2) $F_{X_\alpha}^{X_\beta} \in \mathcal{M}$ for each ordinal $\beta \geq \alpha$.

Remark. Since $0 \in C$ and $X_0 = \emptyset$, property (2) yields $F^{X_\beta} \in \mathcal{M}$ for each $\beta < \mu$.

By Corollaries 3.7 and 3.8, it is necessary for the existence of a marriage of a family F that every μ-test of F is positive.

Of course a μ-test which is not positive is called negative. Using the pigeon-hole principle for stationary sets (I.6.7), it is obvious that, for each negative μ-test $(X_\alpha : \alpha < \mu)$, one of the following properties is satisfied:

(1) There is an ordinal $\alpha < \mu$ such that $F^{X_\alpha} \notin \mathcal{M}$.

(2) $\{\alpha < \mu : \exists i \in I\,(F(i) \subseteq X_\alpha \wedge \forall \beta < \alpha\; F(i) \not\subseteq X_\beta)\}$ is stationary in μ.

(3) $\{\alpha < \mu : \exists \beta > \alpha\; F_{X_\alpha}^{X_\beta} \notin \mathcal{M}\}$ is stationary in μ.

In the introduction we discussed several examples. Now we present a more complicated one.

Example. It is easy to see that

$$J = \{\alpha < \aleph_2 : cf(\alpha) = \aleph_1\}$$

is a stationary subset of ω_2. For each $\alpha < \aleph_2$, denote by α^* the least ordinal number in J which is greater than α.

Claim 1. If $\alpha \in J$ and $\beta < \aleph_1$, then $\alpha \leq \alpha + \beta < \alpha^*$.

Since $\alpha < \alpha^*$ and $cf(\alpha^*) = \aleph_1$, there is a strictly increasing sequence $(\sigma_\xi : \xi < \aleph_1)$ such that $\sigma_0 = \alpha$, $\sigma_\xi < \alpha^*$ for each $\xi < \aleph_1$, and $\sup\{\sigma_\xi : \xi < \aleph_1\} = \alpha^*$. Since $|\beta| \leq \aleph_0$, there is an $\xi < \aleph_1$ such that $\alpha + \beta < \sigma_\xi < \alpha^*$.

Now put $I = \{\alpha + \beta : \alpha \in J \wedge \beta < \aleph_1 \wedge Lim(\beta)\}$.

Claim 2. I is a non-stationary subset of ω_2.

Let C be the union of J and the set of all limits of strictly increasing sequences of elements of J which have countable length. Of course C is a closed unbounded subset of ω_2. Now let us assume that there exists an element $\alpha + \beta \in C \cap I$. Since $cf(\alpha + \beta) = \aleph_0$, it follows that $\alpha + \beta \notin J$. Hence there is a strictly increasing sequence $(\sigma_n : n < \omega)$ of elements of J such that $\sup\{\sigma_n : n < \omega\} = \alpha + \beta$ and $\sigma_0 = \alpha$. But by Claim 1, $\sigma_0 = \alpha < \alpha + \beta < \sigma_1$. This proves $C \cap I = \emptyset$. Therefore I is a non-stationary subset of ω_2.

Now take the family

$$F = (\gamma : \gamma \in I).$$

Let us define a negative $\aleph_2$-test by $X_\alpha = \alpha$ for each $\alpha < \aleph_2$. Since $X_\alpha = \alpha \cup F(\emptyset)$, the sequence $(X_\alpha : \alpha < \aleph_2)$ is an $\aleph_2$-test.

Claim 3. For each $\alpha \in J$, the family $F_\alpha^{\alpha^*}$ has no marriage.

Assume that $F_\alpha^{\alpha^*}$ has a marriage f. For every limit ordinal $\beta < \aleph_1$, there is an ordinal $g(\beta) < \beta$ such that $f(\alpha + \beta) = \alpha + g(\beta)$. By Fodor's Theorem, there is an ordinal $\rho < \omega_1$ such that

$\{\beta < \omega_1 : g(\beta) = \rho\}$

is a stationary subset of ω_1. This contradicts the fact that f is injective.

By Claim 3, the set $\{\alpha < \aleph_2 : \exists \beta > \alpha\, (F_{X_\alpha}^{X_\beta}$ has no marriage$)\}$ is a stationary subset of $\aleph_2$. Hence $(X_\alpha : \alpha < \aleph_2)$ is a negative $\aleph_2$-test of F.

§4. The μ-test criterion

The following example shows that the condition that every μ-test is positive is not sufficient for the existence of a marriage.

Example. Let $I = \{0,1\}$ and $F(0) = \{0\} = F(1)$.
If $(X_\alpha : \alpha < \mu)$ is a μ-test of F, then there is an ordinal $\alpha_0 < \omega_1$ such that $F(i) \subseteq X_{\alpha_0}$ whenever $F(i) \subseteq X_\alpha$ for some $\alpha < \mu$. Define the club C by $C = \{\alpha < \omega_1 : \alpha_0 < \alpha\}$. This club fulfils the conditions (1) and (2) of a positive μ-test. But obviously F has no marriage.

So we are going to extend the condition that every μ-test has to be positive by a further simple property.

Lemma 4.1. If $F \in \mathcal{M}$ and K is critical in $F \setminus A$, then

$|\{i \in I \setminus K : F(i) \subseteq A \cup F(K)\}| \leq |A|$.

Proof. Put $J = \{i \in I : F(i) \subseteq A \cup F(K)\}$. By hypothesis, $F \restriction J$ has a marriage, and clearly $F(J) \subseteq A \cup F(K)$. It follows from Theorem 2.10 that $|J \setminus K| \leq |A| + |K \setminus J|$. But $K \setminus J = \emptyset$, and so $|J \setminus K| \leq |A|$.

Remark. If K_1, K_2 are maximal critical in $F \setminus A$, then, by Theorem 2.7, $F(K_1) \cup A = F(K_2) \cup A$. By Theorem 2.10, it follows that $|K_1 \setminus K_2| = |K_2 \setminus K_1|$, and therefore $|\{i \in I \setminus K_1 : F(i) \subseteq F(K_1) \cup A\}| = |\{i \in I \setminus K_2 : F(i) \subseteq F(K_2) \cup A\}|$. This shows that the following definition is independent of the choice of a maximal critical set: if K is maximal critical in $F \setminus A$, then the cardinal $|\{i \in I \setminus K : F(i) \subseteq F(K) \cup A\}|$ may be called the cardinal demand of A, denoted by $cd(A)$. Furthermore, if L is critical in $F \setminus A$ and $K \supseteq L$ is maximal critical in $F \setminus A$, then $\{i \in I \setminus L : F(i) \subseteq F(L) \cup A\}$ is a subset of $\{i \in I \setminus K : F(i) \subseteq F(K) \cup A\}$, since $(F \upharpoonright K) \setminus A \in \mathcal{M}$. It follows that the statement "for each set A and each K critical in $F \setminus A$, $|\{i \in I \setminus K : F(i) \subseteq F(K) \cup A\}| \leq |A|$" is equivalent to "$cd(A) \leq |A|$ for each set A". - This new condition is more subtle than Hall's condition. To explain this, let K be maximal critical in $F \setminus A$. Hall's condition counts the elements of $D(A) = \{i \in I : F(i) \subseteq A\}$, this new condition however also counts those elements $i \in I \setminus K$ with $F(i) \subseteq F(K) \cup A$ which are not in $D(A)$. The next example will demonstrate this. If the family $F = (F(i) : i \in I)$ is defined by $I = \{\alpha : \omega \leq \alpha < \omega_1\}$ and $F(\alpha) = \alpha$ for each $\alpha \in I$, and if $A = \omega$, then $K = \{\alpha + 1 : \alpha \in I\}$ is maximal critical in $F \setminus A$ and $F(\alpha) \subseteq F(K) \cup A$ for each limit ordinal $\alpha < \omega_1$. Therefore $|A| = \aleph_0 < \aleph_1 = cd(A)$. By Lemma 4.1, F has no marriage.

The following example shows that the condition of Lemma 4.1 is not sufficient for the existence of a marriage.

Example. Put $I = L(\omega_1) = \{\alpha < \omega_1 : \alpha \text{ is a limit ordinal}\}$ and $F(\alpha) = \alpha$ for any $\alpha \in L(\omega_1)$. Now let A be a set and K be critical in $F \setminus A$. We claim that $|\{i \in I \setminus K : F(i) \subseteq A \cup F(K)\}| \leq |A|$. If $|A| = \aleph_1$, then this is obviously true. So let $|A| \leq \aleph_0$. Without loss of generality let A be a subset of ω_1. If $|A| < \aleph_0$, then each member of the family $F \setminus A$ is an infinite set. Thus $K = \emptyset$ by Corollary 2.9. Therefore $\{i \in I \setminus K : F(i) \subseteq A \cup F(K)\} = \{i \in I : F(i) \subseteq A\} = \emptyset$ and so the claim is proved. If $|A| = \aleph_0$, then define

$$\alpha_0 = \sup\{\alpha + 1 : \alpha \in A\}.$$

By Lemma 2.5, there is a set $L \subseteq K$ critical in $F \setminus \alpha_0$ such that

$$F(L) \cup \alpha_0 = F(K) \cup \alpha_0.$$

Again by Corollary 2.9, $L = \emptyset$ and so $F(K) \cup A \subseteq F(K) \cup \alpha_0 = \alpha_0$. Therefore $|\{i \in I \setminus K : F(i) \subseteq A \cup F(K)\}| \leq \aleph_0$, proving the claim.

All these considerations which we already made in the introduction lead us to the basic notion of a λ-positive family.

Definition.
Let λ be a cardinal number. A family F is called λ-positive, if the following conditions hold:

(1) If A is a set such that $|A| \leq \lambda$ and if K is critical in $F \setminus A$, then $|\{i \in I \setminus K : F(i) \subseteq A \cup F(K)\}| \leq |A|$.

(2) If $\mu \leq \lambda$ is an uncountable regular cardinal, then every μ-test of F is positive.

Otherwise F is called λ-negative.

We note that a λ-positive family F is κ-positive for every cardinal $\kappa < \lambda$; further each subfamily of F is λ-positive.

By Corollary 3.7, Corollary 3.8, and Lemma 4.1, we can conclude:

Theorem 4.2. If the family F has a marriage, then F is λ-positive for each cardinal number λ.

We are going to prove that the condition that the family F is λ-positive for each cardinal number λ is also sufficient for the existence of a marriage. But using the notion of the essential size of a family F, denoted by $\|F\|$, we will see that it is enough to check whether F is λ-positive for some cardinal number $\lambda \geq \|F\|$.

Definition. If F is a family, then
$\|F\| = \min\{|A| : \exists K\,(K \text{ critical in } F \setminus A \wedge A \cup F(K) = F(I))\}$
is called the essential size of the family F.

Lemma 4.3. Let F be a family, A be an infinite set, and K be critical in $F \setminus A$ such that $A \cup F(K) = F(I)$. Then the family F has a marriage if and only if there are sets $L \subseteq K$, $B \supseteq A$ such that $|A| = |B|$, L is critical in $F \setminus B$, $B \cup F(L) = F(I)$, and $(F \setminus L) \setminus (F(L) \setminus B)$ has a marriage.

Proof. If g is a marriage of $(F \upharpoonright L) \setminus B$ and h is a marriage of $(F \setminus L) \setminus (F(L) \setminus B)$, then $g \cup h$ is a marriage of F. To prove the converse, let f be a marriage of $(F \upharpoonright K) \setminus A$ and g be a marriage of F. Define $J_0 = I \setminus K$, $J_{n+1} = J_n \cup f^{-1}[g[J_n]]$, and $J = \bigcup\{J_n : n < \omega\}$. Furthermore put $B = A \cup g[J]$ and $L = K \setminus J$. By Lemma 4.1, $|J_0| \leq |A|$. Since A is infinite, it follows that $|J| \leq |A|$, and so $|B| = |A|$. $(F \upharpoonright L) \setminus B$ is critical and

$B \cup F(L) = F(I)$, since $f[L] \cup B = F(K) \cup A = F(I)$. Obviously $g \upharpoonright (I \setminus L)$ is a marriage of $(F \setminus L) \setminus (F(L) \setminus B)$.

Now we state Podewski's main theorem ([P]). Its proof will require the rest of this section.

Theorem 4.4. (K.P. Podewski)
Any family F which is $\|F\|$-positive has a marriage.

It follows from Theorem 4.2 and Theorem 4.4 that a family F has a marriage if and only if F is $\|F\|$-positive.

Let us come to the proof of Theorem 4.4. If $\|F\| = \lambda$, we distinguish the cases that λ is finite, countable or uncountable. The most difficult step - the one for uncountable λ - is divided into two parts: The regular case and the singular case.

Case 1. $\lambda < \aleph_0$.

By the definition of the essential size of a family, there are a set A_0 and a set K_0 critical in $F \setminus A_0$ such that $F(I) = A_0 \cup F(K_0)$. By assumption, $|\{i \in I \setminus K_0 : F(i) \subseteq A_0 \cup F(K_0)\}| \leq |A_0| < \aleph_0$. Let J be a maximal subset of I so that $F \upharpoonright J$ has a marriage and $K_0 \subseteq J$.

Claim. $J = I$.
Let us assume that there is an $i \in I \setminus J$. Then $F(i) \subseteq \operatorname{rng} f$ for each marriage f of $F \upharpoonright J$. By Corollary 2.13, there is a set L critical in $F \upharpoonright J$ such that $F(i) \subseteq F(L)$. Hence $|\emptyset| < |\{i \in I \setminus L : F(i) \subseteq \emptyset \cup F(L)\}|$, contradicting the fact that F is λ-positive. This proves the claim, and thus Theorem 4.4 is proved for each finite cardinal λ.

The following notion of a μ-admissible matching will play an important role in the proof for $\lambda = \aleph_0$ and for singular λ.

Definition. Let $\mu \geq \aleph_0$ be a regular cardinal. If f is a matching in F such that $|f| < \mu$, we call f μ-admissible if $F^D \setminus f$ has a marriage for each set $D \subseteq F(I)$ with $|D| < \mu$.

Case 2. $\lambda = \aleph_0$.

Lemma 4.5. The empty matching in F is $\aleph_0$-admissible.

Proof. Let $D \subseteq F(I)$ be a finite set. Then $\|F^{\bar{D}}\| \leq |D| < \aleph_0$. Since F is $\aleph_0$-positive, F is $|D|$-positive and so the subfamily $F^{\bar{D}}$ is $|D|$-positive. Consequently $F^{\bar{D}}$ has a marriage by Case 1.

Lemma 4.6. If f is an $\aleph_0$-admissible matching in F and $J \subseteq I$ is a finite set, then there is an $\aleph_0$-admissible matching g in F such that $g \supseteq f$ and $J \subseteq \operatorname{dom} g$.

Proof. It suffices to prove Lemma 4.6 for the special case $J = \{i\}$ and $i \notin \operatorname{dom} f$. We distinguish two cases.

Case A. $F(i) \subseteq \overline{\operatorname{rng} f}$.

Let h be a marriage of $F^{\overline{\operatorname{rng} f}} \setminus f$. By assumption, $i \in \operatorname{dom} h$. Put $g := f \cup \{(i, h(i))\}$.

Claim. $F^{\bar{D}} \setminus g$ has a marriage, for each finite set D.

Let us assume that there is a finite set D such that $F^{\bar{D}} \setminus g$ has no marriage. Since f is $\aleph_0$-admissible, the family $(F^{\bar{D}} \setminus f) \setminus \{i\}$ has a marriage. By Theorem 2.11, there is a set L which is critical in $(F^{\bar{D}} \setminus f) \setminus \{i\} \subseteq F \setminus \operatorname{rng} f$ such that $h(i) \subset F(L) \setminus \operatorname{rng} f$. Of course $i \notin L$. Since $F(L) \subseteq \overline{\operatorname{rng} f}$ by the definition of the closure of rng f, L is critical in $F^{\overline{\operatorname{rng} f}} \setminus f$. Therefore $h[L] = F(L) \setminus \operatorname{rng} f$. It follows that $h(i) \in h[L]$ and so $i \in L$, contradicting the fact that $i \notin L$.

Case B. $F(i) \not\subseteq \overline{\operatorname{rng} f}$.

Choose an element $b \in F(i) \setminus \overline{\operatorname{rng} f}$ and define $g := f \cup \{(i,b)\}$.

Claim. $F^{\bar{D}} \setminus g$ has a marriage, for each finite set D.

To get a contradiction, assume that there is a finite set D such that $F^{\bar{D}} \setminus g \notin \mathcal{M}$. Since f is $\aleph_0$-admissible, $(F^{\bar{D}} \setminus f) \setminus \{i\} \in \mathcal{M}$, and so, by Theorem 2.11, there is a set L critical in $(F^{\bar{D}} \setminus f) \setminus \{i\}$ such that $b \in F(L) \setminus \operatorname{rng} f$. So $b \in \overline{\operatorname{rng} f}$, contradicting the choice of b.

To prove Theorem 4.4 for $\lambda = \aleph_0$, let K_0 be critical in $F \setminus A_0$ such that $F(I) = A_0 \cup F(K_0)$ and $\|F\| = |A_0|$. Let f_0 be a marriage of $(F \upharpoonright K_0) \setminus A_0$. Since F is $\aleph_0$-positive, it follows that $|I \setminus K_0| \leq \aleph_0$.

Thus there is an increasing chain $(J_n : n<\omega)$ such that $I\setminus K_0 = \bigcup\{J_n : n<\omega\}$ and $|J_n|<\aleph_0$ for each $n<\omega$. We define recursively a sequence $(g_n : n<\omega)$ of $\aleph_0$-admissible matchings such that

1. $g_n \subseteq g_{n+1}$,
2. $J_n \subseteq \operatorname{dom} g_n$, and
3. If $f_0(i) \in \operatorname{rng} g_n$, then $i \in \operatorname{dom} g_{n+1}$.

Put $g_0 = \emptyset$. By Lemma 4.5, g_0 is an $\aleph_0$-admissible matching. If g_n is defined, put

$$L_{n+1} = J_{n+1} \cup \{i \in K_0 : f_0(i) \in \operatorname{rng} g_n\}.$$

By Lemma 4.6, there is an $\aleph_0$-admissible matching $g_{n+1} \supseteq g_n$ such that $L_{n+1} \subseteq \operatorname{dom} g_{n+1}$. If $g = \bigcup\{g_n : n<\omega\}$, then $h = g \cup f_0 \restriction (K_0 \setminus \operatorname{dom} g)$ is a marriage of F.

<u>Case 3</u>. $\lambda > \aleph_0$.

We distinguish the two cases that λ is regular or λ is singular.

<u>The regular case</u>

Since the regular case is fairly complicated, we give a short outline of the proof.

Let K_0 be critical in $F\setminus A_0$ such that $F(I) = A_0 \cup F(K_0)$ and $\|F\| = |A_0| = \lambda$, where λ is regular. Since we suppose that F is λ-positive, it follows that $|I\setminus K_0| \leq \lambda$. First we claim that there exists a marriage h_0 of $F\restriction(I\setminus K_0)$ such that $\operatorname{rng} h_0$ is not too scattered. To be precise, we let $H_\lambda(B) = \bigcup\{\overline{D} : D \subset [B]^{<\lambda}\}$ for any set B and require that there is a set $B_1 \supseteq A_0$ with $|B_1| \leq \lambda$ such that $\operatorname{rng} h_0 \subseteq H_\lambda(B_1)$. We want to describe how this can be arranged. Let B be a set of women. A man i is called <u>B-constrained</u> if there exists a set $D \subset [B]^{<\lambda}$ such that $F(i) \subseteq \overline{D}$. The man i is said to be <u>B-free</u> if $F(i) \cap B \not\subseteq \overline{D}$ for each $D \in [B]^{<\lambda}$. Lemma 4.7 states that there is a set $B_1 \supseteq A_0$ with $|B_1| \leq \lambda$ such that every man $i \in I\setminus K_0$ is B_1-constrained or B_1-free. By Lemma 4.9, the family $F\restriction\{i \in I : i \text{ is } B_1\text{-constrained}\}$ has a marriage. If a man $i \in I\setminus K_0$ is B_1-free, then he knows many women of $H_\lambda(B_1)$, and so, by Lemma 4.10, we can arrange a marriage g_0 of $F\restriction\{i \in I\setminus K_0 : i \text{ is } B_1\text{-constrained}\}$ which leaves for him enough women unmarried. Therefore it is possible to extend g_0 to a marriage h_0 of $F\restriction(I\setminus K_0)$ such that $\operatorname{rng} h_0 \subseteq H_\lambda(B_1)$. This proves our first claim.

Now let f_0 be a marriage of $F\restriction K_0$. Unfortunately the choice function $f_0 \cup h_0$ is not necessarily injective. Hence define $J_1 = (I\setminus K_0) \cup \{i \in K_0 : f_0(i) \in H_\lambda(B_1)\}$. The proof of Lemma 4.8 shows that $|\{i \in J_1 : i$ is not B_1-constrained$\}| \leq \lambda$. By Lemma 4.7, there is a set $B_2 \supseteq B_1$ with $|B_2| \leq \lambda$ such that every man $i \in J_1$ is B_2-constrained or B_2-free. Again we can show that $F \upharpoonright J_1$ has a marriage h_1 with $\text{rng}\, h_1 \subseteq H_\lambda(B_2)$. Then $f_0 \upharpoonright (I \setminus J_1) \cup h_1$ is a choice function which is perhaps not injective. So we iterate this process λ times and define chains $(J_\alpha : \alpha < \lambda)$ and $(B_\alpha : \alpha < \lambda)$ such that $J_0 = I \setminus K_0$, $B_0 = A_0$, $|B_\alpha| \leq \lambda$, each man $i \in J_\alpha$ is $B_{\alpha+1}$-constrained or $B_{\alpha+1}$-free, and $J_{\alpha+1} = (I \setminus K_0) \cup \{i \in K_0 : f_0(i) \in H_\lambda(B_{\alpha+1})\}$. Put $B = \bigcup\{B_\alpha : \alpha < \lambda\}$ and $J = \bigcup\{J_\alpha : \alpha < \lambda\}$. Then $[B]^{<\lambda} = \bigcup\{[B_\alpha]^{<\lambda} : \alpha < \lambda\}$, and so each man of J is B-constrained or B-free. By Lemma 4.10, there exists a marriage h of $F \upharpoonright J$ with $\text{rng}\, h \subseteq H_\lambda(B)$. Since $H_\lambda(B) = \bigcup\{H_\lambda(B_\alpha) : \alpha < \lambda\}$, it follows that $\{i \in K_0 : f_0(i) \in H_\lambda(B)\} = \bigcup\{\{i \in K_0 : f_0(i) \in H_\lambda(B_\alpha)\} : \alpha < \lambda\} \subseteq$ $= \bigcup\{J_{\alpha+1} : \alpha < \lambda\} \subseteq J$. Therefore $f_0 \upharpoonright (I \setminus J) \cup h$ is a marriage of F.

<u>Definition</u>. For each family $F = (F(i) : i \in I)$ and each set B, let
$I_c(B) = \{i \in I : \exists D \in [B]^{<\lambda}\ F(i) \subseteq \overline{D}\}$ and
$I_f(B) = \{i \in I : \forall D \in [B]^{<\lambda}\ F(i) \cap B \not\subseteq \overline{D}\}$. $I_c(B)$, $I_f(B)$ is called the set of B-constrained men, B-free men respectively.

<u>Definition</u>. Let $J \subseteq I$ and $B \subseteq F(I)$. An ordered pair (J,B) is called a <u>λ-constructive pair</u> if $J \subseteq I_c(B) \cup I_f(B)$, $|J \cap I_f(B)| \leq \lambda$, and $|B| \leq \lambda$.

If (J,B) is a λ-constructive pair, then $(J \cup I_c(B), B)$ and $(J \setminus I_c(B), B)$ also are λ-constructive pairs.

<u>Lemma 4.7</u>. For any set $J \subseteq I$ such that $|J| \leq \lambda$ and any set $A \subseteq F(I)$ with $|A| \leq \lambda$, there exists a set $B \supseteq A$ such that (J,B) is a λ-constructive pair.

<u>Proof</u>. Let $(i_\alpha : \alpha < \lambda)$ be an enumeration of J such that $|\{\alpha : i = i_\alpha\}| = \lambda$ for each $i \in J$. Let $(A_\alpha : \alpha < \lambda)$ be a continuous chain such that $A = \bigcup\{A_\alpha : \alpha < \lambda\}$ and $|A_\alpha| < \lambda$ for all $\alpha < \lambda$. We define recursively a continuous chain $(B_\alpha : \alpha < \lambda)$ such that $|B_\alpha| < \lambda$ for each $\alpha < \lambda$. Put $B_0 = \emptyset$ and let B_β be defined for all $\beta < \alpha$. If α is a limit ordinal, then put $B_\alpha = \bigcup\{B_\beta : \beta < \alpha\}$. Otherwise $\alpha = \beta + 1$ for some β. If $F(i_\beta) \subseteq \overline{B}_\beta$, then we put $B_{\beta+1} = B_\beta \cup A_\alpha$. If $F(i_\beta) \not\subseteq \overline{B}_\beta$, then choose an element b of $F(i_\beta) \setminus \overline{B}_\beta$ and define $B_{\beta+1} = B_\beta \cup A_\alpha \cup \{b\}$. If we take $B = \bigcup\{B_\alpha : \alpha < \lambda\}$, then (J,B) is a λ-constructive pair. To prove this,

let $i \in J \setminus I_c(B)$. Then $F(i) \not\subseteq \overline{B}_\alpha$ for any $\alpha < \lambda$. For each $\alpha < \lambda$, there is an ordinal $\gamma \geq \alpha$ such that $i = i_\gamma$. Therefore, by construction, $F(i_\gamma) \cap B_{\gamma+1} \not\subseteq \overline{B}_\gamma$ and so all the more $F(i) \cap B \not\subseteq \overline{B}_\alpha$. Thus $F(i) \cap B \not\subseteq \overline{D}$ for all $D \in [B]^{<\lambda}$. This proves that $i \in I_f(B)$.

<u>Lemma 4.8</u>. There exists a λ-constructive pair (J,B) such that $(I \setminus K_0) \cup I_c(B) \subseteq J$, $B \supseteq A_0$, and $f_0[K_0 \setminus J] \cap H_\lambda(B) = \emptyset$.

<u>Proof</u>. We define recursively chains $(B_\alpha : \alpha < \lambda)$ and $(J_\alpha : \alpha < \lambda)$ such that, for any $\alpha > 0$, $I \setminus K_0 \subseteq J_\alpha$, $|J_\alpha| \leq \lambda$, $A_0 \subseteq B_\alpha$, $|B_\alpha| \leq \lambda$, and

(a) $(J_\alpha, B_{\alpha+1})$ is a λ-constructive pair.

(b) If $i \in K_0 \setminus I_c(B_\alpha)$ and $f_0(i) \subset H_\lambda(B_\alpha)$, then $i \in J_{\alpha+1}$.

Let us assume that such chains are defined. Put

$$B = \cup\{B_\alpha : \alpha < \lambda\} \text{ and}$$

$$J = \cup\{J_\alpha : \alpha < \lambda\} \cup I_c(B).$$

In order to prove that (J,B) is a λ-constructive pair, let $i \in J \setminus I_c(B)$. Suppose for contradiction that there is a set $D \in [B]^{<\lambda}$ such that $F(i) \cap B \subseteq \overline{D}$. Then there is an $\alpha < \lambda$ such that $i \in J_\alpha \setminus I_c(B_{\alpha+1})$ and $F(i) \cap B_{\alpha+1} \subseteq \overline{D}$. It follows that $(J_\alpha, B_{\alpha+1})$ is not a λ-constructive pair, contradicting property (a). Therefore (J,B) is λ-constructive.

To prove the second part of the lemma, assume for contradiction that there is an $i \in K_0 \setminus J$ such that $f_0(i) \in H_\lambda(B)$. Since λ is regular, there are an ordinal α and a set $D \in [B_\alpha]^{<\lambda}$ such that $f_0(i) \in \overline{D}$. Since $i \in I \setminus I_c(B_\alpha)$, it follows by property (b) that $i \in J_{\alpha+1} \subseteq J$, which contradicts the fact that $i \in K_0 \setminus J$. This proves $f_0[K_0 \setminus J] \cap H_\lambda(B) = \emptyset$.

Now let us construct the sequences $(J_\alpha : \alpha < \lambda)$, $(B_\alpha : \alpha < \lambda)$. Put $J_0 = I \setminus K_0$ and $B_0 = A_0$. If α is a limit ordinal, put $J_\alpha = \cup\{J_\beta : \beta < \alpha\}$ and $B_\alpha = \cup\{B_\beta : \beta < \alpha\}$. If $\alpha = \beta + 1$, then let us assume that $|J_\beta| \leq \lambda$. Then, by Lemma 4.7, there exists a set $B_\alpha \supseteq B_\beta$ such that (J_β, B_α) is λ-constructive. Define

$$J_\alpha = J_\beta \cup \{i \in K_0 \setminus I_c(B_\alpha) : f_0(i) \subset H_\lambda(B_\alpha)\}.$$

We have to prove the

<u>Claim</u>. $|J_{\beta+1}| \leq \lambda$.

Let $(D_\xi : \xi < \lambda)$ be a continuous chain such that $|D_\xi| < \lambda$ for all $\xi < \lambda$ and $B_\alpha = \cup\{D_\xi : \xi < \lambda\}$. For each $\xi < \lambda$, put

$$J_{D_\xi} = \{i \in K_0 \setminus I_c(B_\alpha) : f_0(i) \in \overline{D}_\xi\}.$$

Since $J_{\beta+1} = J_\beta \cup \bigcup\{J_{D_\xi} : \xi < \lambda\}$, it is enough to show that $|J_{D_\xi}| \leq \lambda$ for each $\xi < \lambda$. Let L be critical in $F \setminus D_\xi$ such that $\overline{D}_\xi = D_\xi \cup F(L)$. If $i \in L$, then $F(i) \subseteq \overline{D}_\xi$ and so $i \in I_c(B_\alpha)$. It follows that no element of L is an element of J_{D_ξ}. Therefore $f_0 \restriction L \cap K_0$ is a marriage of $(F \restriction L \cap K_0) \setminus f_0[J_{D_\xi}]$. Furthermore $F(L \cap K_0) \cup f_0[J_{D_\xi}] \subseteq F(L) \cup D_\xi = \overline{D}_\xi$.
By Theorem 2.10,

$$|(L \cap K_0) \setminus L| + |f_0[J_{D_\xi}]| \leq |L \setminus (L \cap K_0)| + |D_\xi|$$

and so, since $|I \setminus K_0| \leq \lambda$,

$$|J_{D_\xi}| = |f_0[J_{D_\xi}]| \leq |L \setminus K_0| + |D_\xi| \leq \lambda.$$

This proves the claim and completes the proof of Lemma 4.8.

To prove the important Lemma 4.10 we need the following observation.

<u>Lemma 4.9</u>. If $|B| \leq \lambda$, then $F \restriction I_c(B)$ has a marriage.

<u>Proof</u>. Let $(A_\alpha : \alpha < \lambda)$ be a continuous chain such that $|A_\alpha| < \lambda$ for all $\alpha < \lambda$ and $B = \bigcup\{A_\alpha : \alpha < \lambda\}$. For $\alpha < \lambda$, put $X_\alpha = \bigcup\{\overline{A}_{\beta+1} : \beta < \alpha\}$. By Lemma 3.5, $(X_\alpha : \alpha < \lambda)$ is a λ-test. Since F is λ-positive, the test $(X_\alpha : \alpha < \lambda)$ is λ-positive. By the definition of a λ-positive test, there is a club C in λ with $0 \in C$ such that, for all $\alpha \in C$, the following holds:

(1) If $F(i) \subseteq X_\alpha$, then there is an ordinal $\beta < \alpha$ such that $F(i) \subseteq X_\beta$.

(2) For any $\beta \geq \alpha$, the family $F_{X_\alpha}^{X_\beta}$ has a marriage.

Let $(s(\alpha) : \alpha < \lambda)$ be an enumeration of C such that $s(\alpha) < s(\beta)$ for any α and β with $\alpha < \beta < \lambda$. By property (2), the family $F_{X_{s(\alpha)}}^{X_{s(\alpha+1)}}$ has a marriage h_α. Since $\operatorname{rng} h_\alpha \subseteq X_{s(\alpha+1)} \setminus X_{s(\alpha)}$, the function $h = \bigcup\{h_\alpha : \alpha < \lambda\}$ is a matching in F.

<u>Claim</u>. $I_c(B) = \operatorname{dom} h$.

Let $i \in I_c(B)$. Then there is a set $D \in [B]^{<\lambda}$ such that $F(i) \subseteq \overline{D}$. Since λ is regular, there is an ordinal $\alpha < \lambda$ such that $D \subseteq A_{\alpha+1}$. Therefore $F(i) \subseteq X_{\alpha+1}$. It follows that

$$\Delta = \{s(\beta) : F(i) \not\subseteq X_{s(\beta)}\}$$

is a bounded nonempty subset of λ. Since C is a closed subset of λ, there is a $\gamma < \lambda$ such that $s(\gamma) = \sup \Delta$. By property (1), $F(i) \not\subseteq X_{s(\gamma)}$.

But $F(i) \subseteq X_{s(\gamma+1)}$, and therefore $i \in \operatorname{dom} h_\gamma \subseteq \operatorname{dom} h$.

For the proof of the converse inclusion, let $i \in \operatorname{dom} h$. Then there is an ordinal $\alpha < \lambda$ such that $i \in \operatorname{dom} h_\alpha$. Therefore, by Lemma 3.5,

$$F(i) \subseteq X_{s(\alpha+1)} \subseteq \overline{X}_{s(\alpha+1)} = \overline{A}_{s(\alpha+1)},$$

and so $i \in I_c(B)$.

<u>Lemma 4.10</u>. If $|J| \leq \lambda$ and (J,B) is a λ-constructive pair, then there is a matching h in the family F such that $\operatorname{dom} h = J \cup I_c(B)$ and $\operatorname{rng} h \subseteq H_\lambda(B)$. Moreover, if f is a λ-admissible matching with $\operatorname{dom} f \subseteq J \cup I_c(B)$ and $\operatorname{rng} f \subseteq B$, then there is a matching h such that $\operatorname{dom} h = J \cup I_c(B)$, $\operatorname{rng} h \subseteq H_\lambda(B)$, and $f \subseteq h$.

<u>Proof</u>. By Lemma 4.9, the family $F \restriction I_c(B)$ has a marriage f^*. We want to construct a marriage of $F \restriction J \setminus I_c(B)$. Since (J,B) is a λ-constructive pair, every B-free man i of J knows many women of B, i.e. $(F(i) \cap B) \setminus \overline{D} \neq \emptyset$ for any $D \in [B]^{<\lambda}$. We show that, by Lemma 3.3, some acquaintances of the men in $I_c(B)$ can be set free to be married with men in $J \cap I_f(B)$.

Let $(D_\alpha : \alpha < \lambda)$ be a continuous chain such that $D_1 = \operatorname{rng} f$, $B = \bigcup\{D_\alpha : \alpha < \lambda\}$, and $|D_\alpha| < \lambda$ for any $\alpha < \lambda$. By Lemma 3.5, the sequence $(X_\alpha : \alpha < \lambda)$, defined by $X_\alpha = \bigcup\{\overline{D}_{\beta+1} : \beta < \alpha\}$, is a λ-test. If $F(i) \subseteq X_\alpha$, then $F(i) \subseteq X_\alpha \subseteq \overline{X}_\alpha = \overline{D}_\alpha$ and so

$$\{i \in I : \exists \alpha < \lambda \ F(i) \subseteq X_\alpha\} \subseteq I_c(B).$$

By Lemma 3.6, there is a club C in λ with $0 \in C$ such that, for all $\alpha \in C$, the following holds:

(*) If $f^*(i) \in X_\alpha$, then $F(i) \subseteq X_\alpha$.

Now let $(i_\alpha : \alpha < \nu)$ be an enumeration of $J \cap I_f(B)$. Of course $\nu \leq \lambda$. We define recursively a strictly increasing function $s : \nu \to C$ and a marriage g of $F \restriction \{i_\alpha : \alpha < \nu\}$ such that

(1) $0 < s(0)$ and

(2) $g(i_\alpha) \in (F(i_\alpha) \cap X_{s(\alpha+1)}) \setminus \overline{X}_{s(\alpha)}$.

Choose $s(0) \in C \setminus \{0\}$ and assume that $s(\alpha)$ is defined for each $\alpha < \beta$. Let $\beta = \alpha + 1$ be a successor ordinal. Since $i_\alpha \in J \cap I_f(B)$, it follows that

$$F(i_\alpha) \cap B \not\subseteq \overline{D}_{s(\alpha)} = \overline{X}_{s(\alpha)}.$$

Choose an element $g(i_\alpha) \in (F(i_\alpha) \cap B) \setminus \overline{X}_{s(\alpha)}$ and an element $s(\alpha+1) \in C$ such that $s(\alpha) < s(\alpha+1)$ and $g(i_\alpha) \in X_{s(\alpha+1)}$. If β is a limit ordinal, then put $s(\beta) = \sup\{s(\alpha) : \alpha<\beta\}$. Clearly $s(\beta)\in C$.

Now we partition $I_c(B) = \operatorname{dom} f^*$ by

$K_\alpha = \{i\in I_c(B) : f^*(i)\in X_{s(\alpha+1)}\setminus X_{s(\alpha)}\}$ for $\alpha<\nu$ and

$K_\nu = \{i\in I_c(B) : f^*(i)\notin \bigcup\{X_{s(\alpha+1)}\setminus X_{s(\alpha)} : \alpha<\nu\}\}$.

Since $X_{s(\alpha+1)}\subseteq \overline{D}_{s(\alpha+1)}$, it follows that $\operatorname{dom} F_{X_{s(\alpha)}}^{X_{s(\alpha+1)}} \subseteq I_c(B)$. By property (*), $f^* \restriction K_\alpha$ is a marriage of $F_{X_{s(\alpha)}}^{X_{s(\alpha+1)}}$. By Lemma 3.3, the family $F_{X_{s(\alpha)}}^{X_{s(\alpha+1)}}$ has a marriage h_α such that $g(i_\alpha)\notin \operatorname{rng} h_\alpha$. Since $s(0)>0$, it follows that $\operatorname{rng} f\subseteq D_{s(0)}\subseteq X_{s(0)}$ and, since f is λ-admissible, the family $F^{X_{s(0)}}\setminus f$, which is a subfamily of $F^{D_{s(0)}}\setminus f$, has a marriage e. Put

$L = \{i\in I : F(i)\subseteq X_{s(0)}\}$ and

$h = f\cup e\cup g\cup \bigcup\{h_\alpha : \alpha<\nu\}\cup f^*\restriction (K_\nu\setminus (L\cup \operatorname{dom} f))$.

If $f^*(i)\in X_{s(0)}$, then, by property (*), $i\in L$. Therefore h is a marriage of $F\restriction J\cup I_c(B)$ such that $\operatorname{rng} h\subseteq H_\lambda(B)$ and $f\subseteq h$. This proves Lemma 4.10.

Now it is easy to prove Theorem 4.4 for uncountable regular cardinals λ. By Lemma 4.8, there is a λ-constructive pair (J,B) such that $I\setminus K_0\subseteq J$ and $f_0[K_0\setminus J]\cap H_\lambda(B) = \emptyset$. By Lemma 4.10, there is a marriage h of $F\restriction J$ such that $\operatorname{rng} h\subseteq H_\lambda(B)$. Therefore $f\restriction (K_0\setminus J)\cup h$ is a marriage of F.

The singular case

Remember that F is a λ-positive family and $\|F\| = \lambda$. Let A_0 be a set and K_0 be critical in $F\setminus A_0$ such that $F(I) = A_0\cup F(K_0)$ and $\|F\| = |A_0| = \lambda$. Since F is λ-positive, it follows that $|I\setminus K_0|\leq\lambda$. Let f_0 be a marriage of $F\restriction K_0$. Put $\nu = \operatorname{cf}(\lambda)$. Since λ is singular, there is a strictly increasing sequence $(\nu_\alpha : \alpha<\lambda)$ of cardinals ν_α such that $\nu<\nu_\alpha<\lambda$ for each $\alpha<\nu$ and $\lambda = \sup\{\nu_\alpha : \alpha<\lambda\}$.

Theorem 4.11.

Let $\mu<\lambda$ be an uncountable regular cardinal.

(a) The empty matching in F is μ-admissible.

(b) If f is a μ-admissible matching in F and $J\subseteq I$ such that $|J|<\mu$, then there exists a μ-admissible matching $g\supseteq f$ such that $J\subseteq \operatorname{dom} g$.

First of all we show that, for the proof of the singular case, it suffices to prove Theorem 4.11. To see this, choose a continuous chain $(J_\alpha : \alpha < \nu)$ such that $I \setminus K_0 = \cup\{J_\alpha : \alpha < \nu\}$ and $|J_\alpha| \leq \nu_\alpha$ for all $\alpha < \nu$. For each ordinal $\alpha \leq \nu$, we define by induction a sequence $(g_n^\alpha : n < \omega)$ of matchings such that

1. $g_n^\nu = f_0$.
2. If $\alpha \leq \nu$, then $g_n^\alpha \subseteq g_{n+1}^\alpha$.
3. For any $\alpha < \nu$, the matching g_n^α is a ν_α^+-admissible matching in F such that $J_\alpha \subseteq \mathrm{dom}\, g_n^\alpha$.
4. If $\alpha < \beta \leq \nu$ and $g_n^\beta(i) \in \mathrm{rng}\, g_n^\alpha$, then $i \in \mathrm{dom}\, g_{n+1}^\alpha$.

Put $g_n^\nu = f_0$ for all $n < \omega$. By Theorem 4.11, the empty matching is ν_α^+-admissible. Since $|J_\alpha| < \nu_\alpha^+$, there is, by Theorem 4.11, a ν_α^+-admissible matching g_0^α such that $J_\alpha \subseteq \mathrm{dom}\, g_0^\alpha$ for any $\alpha < \nu$. Assume that g_n^α is defined for each $\alpha < \nu$. Put $J_\alpha^n := \{i \in I : \exists \beta \leq \nu (\alpha < \beta \wedge g_n^\beta(i) \in \mathrm{rng}\, g_n^\alpha)\}$. Then $|J_\alpha^n| \leq \nu \cdot \nu_\alpha = \nu_\alpha < \nu_\alpha^+$. By Theorem 4.11, there exists a ν_α^+-admissible matching $g_{n+1}^\alpha \supseteq g_n^\alpha$ such that $J_\alpha \cup J_\alpha^n \subseteq \mathrm{dom}\, g_{n+1}^\alpha$.

Put $g^\alpha := \cup\{g_n^\alpha : n < \omega\}$ for all $\alpha \leq \nu$. Then

(1) $g^\nu = f_0$,

(2) $J_\alpha \subseteq \mathrm{dom}\, g^\alpha$, and

(3) If $\alpha < \beta \leq \nu$ and $g^\beta(i) \in \mathrm{rng}\, g^\alpha$, then $i \in \mathrm{dom}\, g^\alpha$.

By property (3), it is clear that $g^\beta(i) \notin \mathrm{rng}\, g^\alpha$ if $\alpha < \beta \leq \nu$ and $i \in \mathrm{dom}\, g^\beta \setminus \mathrm{dom}\, g^\alpha$. For each $\alpha \leq \nu$, put

$L_\alpha := \mathrm{dom}\, g^\alpha \setminus \cup\{\mathrm{dom}\, g^\gamma : \gamma < \alpha\}$.

By construction, $I = \cup\{L_\alpha : \alpha \leq \nu\}$. For any $\alpha \leq \nu$, the function $g^\alpha \restriction L_\alpha$ is a marriage of $F \restriction L_\alpha$ such that $g^\alpha[L_\alpha] \cap g^\beta[L_\beta] = \emptyset$, if $\alpha < \beta \leq \nu$. Therefore $h = \cup\{g^\alpha \restriction L_\alpha : \alpha \leq \nu\}$ is a marriage of F.

Part (a) of Theorem 4.11 is easy to prove: If $\mu < \lambda$ is an uncountable regular cardinal and if $D \in [\Gamma(I)]^{<\mu}$, then the μ-test $(X_\alpha : \alpha < \mu)$, defined by $X_\alpha = \bar{\bar{D}}$, is μ-positive. Therefore the family F^D has a marriage. To prove part (b) of Theorem 4.11, it suffices to prove the following lemma.

Lemma 4.12.

If $\mu < \lambda$ is an uncountable regular cardinal, f is a μ-admissible matching in F, and $J \subseteq I$ is a set with $\mathrm{dom}\, f \subseteq J$ and $|J| < \mu$, then there exists a set $B \supseteq \mathrm{rng}\, f$ with $|B| \leq \mu$ and the following properties:

(a) (J,B) is a μ-constructive pair.

(b) For any $D\in[B]^{<\mu}$, there exists a set $A\in[B]^{<\mu}$ with $D\subseteq A$ such that, for any $C\in[F(I)]^{<\mu}$ with $C\supseteq A$, the family $F_{\overline{A}}^{\overline{C}}$ has a marriage.

Before proving Lemma 4.12, let us show that Lemma 4.12 implies part (b) of Theorem 4.11. Let f be a μ-admissible matching in F and $J\in[I]^{<\mu}$. Without loss of generality we can assume $\operatorname{dom} f\subseteq J$. By Lemma 4.12, there is a set $B\supseteq \operatorname{rng} f$ with $|B|\leq\mu$ such that (J,B) is a μ-constructive pair with property (b) of Lemma 4.12. Lemma 4.10 yields a marriage h of $F\restriction J\cup I_c(B)$ such that $f\subseteq h$ and $\operatorname{rng} h\subseteq H_\lambda(B)$. It remains to show that the matching $g = h\restriction J$ is a μ-admissible matching. Obviously $|g|<\mu$. Since $\operatorname{rng} h\subseteq H_\lambda(B)$, there exists a set $D\in[B]^{<\mu}$ such that $\operatorname{rng} g\subseteq\overline{D}$. Property (b) of Lemma 4.12 yields a set $A\in[B]^{<\mu}$ with $D\subseteq A$ such that $F_{\overline{A}}^{\overline{C}}$ has a marriage for each set $C\supseteq A$ with $|C|<\mu$. Now if $E\in[F(I)]^{<\mu}$, then the family $F_{\overline{A}}^{\overline{E\cup A}}$ has a marriage g'. But $\operatorname{rng} g\subseteq\overline{A}$, since $D\subseteq A$, and so $h\restriction\{i\in I\setminus\operatorname{dom} g : F(i)\subseteq\overline{A}\}\cup g'$ is a marriage of $F^{\overline{E\cup A}}\setminus g$. This proves that $F^{\overline{E}}\setminus g\in\mathcal{M}$ for each set E with $|E|<\mu$.

It remains to prove Lemma 4.12. We will do this with help of Lemma 4.13.

Lemma 4.13. If μ is an uncountable regular cardinal, then, for any set D with $|D|<\mu$, there is a set $A\supseteq D$ with $|A|<\mu$ such that the family $F_{\overline{A}}^{\overline{C}}$ has a marriage for each set $C\supseteq A$ with $|C|<\mu$.

Proof. In order to get a contradiction let us assume that there is a set D with $|D|<\mu$ such that, for every set $A\supseteq D$ with $|A|<\mu$, there is a set $C\supseteq A$ with $|C|<\mu$ such that the family $F_{\overline{A}}^{\overline{C}}$ has no marriage. By recursion, we define a continuous chain $(A_\alpha:\alpha<\mu)$ as follows: Let $A_0=\emptyset$, $A_1 = D$ and $A_{\alpha+1}\supseteq A_\alpha$ be a set with $|A_{\alpha+1}|<\mu$ such that

(*) $\quad F_{\overline{A}_\alpha}^{\overline{A}_{\alpha+1}}$ has no marriage.

Define a μ-test $(X_\alpha:\alpha<\mu)$ by $X_\alpha = \bigcup\{\overline{A}_{\beta+1}:\beta<\alpha\}$. Since F is μ-positive, the μ-test $(X_\alpha:\alpha<\mu)$ is μ-positive, too. It follows that there is a club C in μ such that $F_{X_\alpha}^{X_{\alpha+1}}$ has a marriage for each $\alpha\in C$. Choose an element α of C such that $\alpha\geq 1$. By Lemma 3.4, the family $F_{\overline{X}_\alpha}^{\overline{X}_{\alpha+1}}$ also has a marriage. By Lemma 3.5, $F_{\overline{A}_\alpha}^{\overline{A}_{\alpha+1}} = F_{\overline{X}_\alpha}^{X_{\alpha+1}}$ and so $F_{\overline{A}_\alpha}^{\overline{A}_{\alpha+1}} \in \mathcal{M}$, contradicting (*).

Proof of Lemma 4.12. Let $(i_\alpha : \alpha < \mu)$ be an enumeration of J such that $|\{\alpha : i = i_\alpha\}| = \mu$ for each $i \in I$. We define by recursion a chain $(B_\alpha : \alpha < \mu)$. Put $B_0 = \operatorname{rng} f$ and $B_\alpha = \bigcup\{B_\beta : \beta < \alpha\}$ for any limit ordinal α. If $\alpha = \beta + 1$ is a successor ordinal, then there is, by Lemma 4.13, a set $A_\alpha \supseteq B_\beta$ with $|A_\alpha| < \mu$ such that $F_{\overline{A}_\alpha}^{\overline{C}} \in \mathcal{M}$ for each set $C \supseteq A_\alpha$ with $|C| < \mu$. If $F(i_\alpha) \subseteq \overline{A}_\alpha$, then define $B_\alpha = A_\alpha$. Otherwise choose an element b of $F(i_\alpha) \setminus \overline{A}_\alpha$ and put $B_\alpha = A_\alpha \cup \{b\}$. Finally let $B = \bigcup\{B_\alpha : \alpha < \mu\}$.

The proof of Lemma 4.7 shows that property (a) of Lemma 4.12 holds. To prove property (b), let $D \in [B]^{<\mu}$. Then there is an ordinal $\beta < \mu$ such that $D \subseteq B_\beta$. Take $A = A_{\beta+1}$. Then, by the choice of $A_{\beta+1}$, $D \subseteq A$ and $F_{\overline{A}}^{\overline{C}} \in \mathcal{M}$ for each set $C \supseteq A$ with $|C| < \mu$.

We point out that Theorem 4.4 yields a compactness theorem whenever $\|F\|$ is singular. To give some details, let $F = (F(i) : i \in I)$ be a family such that $\|F\| = \lambda$ is singular and suppose

(i) There exists a set $L \subseteq I$ such that $F \restriction L \in \mathcal{M}$ and $|I \setminus L| \le \lambda$ and

(ii) $H \in \mathcal{M}$ for any subfamily H of F with $\|H\| < \lambda$.

Then the family F has a marriage.

To prove this we have to check that F is λ-positive. Let $\mu < \lambda$ be an uncountable regular cardinal and $(X_\alpha : \alpha < \mu)$ be a μ-test of F. Then, for each $\alpha < \mu$, there are a set A_α with $|A_\alpha| < \mu$ and a set K_α critical in $F \setminus A_\alpha$ such that $X_\alpha = F(K_\alpha) \cup \overline{A}_\alpha$. If $A = \bigcup\{A_\alpha : \alpha < \mu\}$, then $|A| \le \mu$ and therefore $\|F^{\overline{A}}\| \le \mu$. By assumption, $F^{\overline{A}} \in \mathcal{M}$. Since $\{i \in I : \exists \alpha < \mu (F(i) \subseteq X_\alpha)\}$ is a subset of $\operatorname{dom} F^{\overline{A}}$, we conclude, by Lemma 3.6 and Corollary 3.8, that $(X_\alpha : \alpha < \mu)$ is positive. This proves that every μ-test of F is positive. Now let K be critical in $F \setminus A$. It remains to show that (1) $|\{i \in I \setminus K : F(i) \subseteq F(K) \cup \overline{A}\}| \le |A|$.
If $|A| < \lambda$, then $\|F^{\overline{A}}\| < \lambda$. By property (ii), $F^{\overline{A}} \in \mathcal{M}$ and, by Lemma 3.1, $\{i \in I : F(i) \subseteq F(K) \cup A\} \subseteq \operatorname{dom} F^{\overline{A}}$. Therefore, by Lemma 4.1, property (1) is fulfilled. If $|A| = \lambda$, then put $J = \{i \in I : F(i) \subseteq F(K) \cup A\} \cap L$. By Theorem 2.10, $|J \setminus K| \le |A| + |K \setminus J|$. Since $\{i \in I \setminus K : F(i) \subseteq F(K) \cup A\} \subseteq (J \setminus K) \cup (I \setminus L)$ and $K \setminus J \subseteq I \setminus L$, it follows that $|\{i \in I \setminus K : F(i) \subseteq F(K) \cup A\}| \le |J \setminus K| + |I \setminus L| \le |A| + |K \setminus J| + |I \setminus L| \le$
$\le |A| + |I \setminus L| + |I \setminus L| \le \lambda = |A|$. This proves (1).

By the μ-test criterion, the existence of a marriage of a family F depends on the existence of marriages of suitable families $F_{X_\alpha}^{X_\beta}$. To avoid this we introduce the notion of a λ-blocked family.

Definition.
If λ is a cardinal and F is a family, then F is called λ-unblocked if the following holds.

(i) For any set A with $|A| \leq \lambda$ and any set K critical in $F \setminus A$, $|\{i \in I \setminus K : F(i) \subseteq A \cup F(K)\}| \leq |A|$.

(ii) For each uncountable regular cardinal $\mu \leq \lambda$ and each μ-test $(X_\alpha : \alpha < \mu)$, there is a club C in μ with $0 \in C$ such that, for all $\alpha \in C$, the following holds:

(a) For any $\beta \geq \alpha$, the family $F_{X_\alpha}^{X_\beta}$ is ρ-unblocked for all $\rho < \mu$.

(b) If $F(i) \subseteq X_\alpha$, then there is an ordinal $\beta < \alpha$ such that $F(i) \subseteq X_\beta$.

Theorem 4.14. A family F is λ-positive if and only if F is λ-unblocked.

Proof. We prove Theorem 4.14 by induction on λ. It is enough to prove that, for every uncountable regular cardinal $\mu \leq \lambda$ and every μ-test $(X_\alpha : \alpha < \mu)$, the following two statements are equivalent for any ordinals α and β such that $\alpha \leq \beta < \mu$:

(i) $F_{X_\alpha}^{X_\beta}$ has a marriage.

(ii) $F_{X_\alpha}^{X_\beta}$ is ρ-unblocked for each cardinal $\rho < \mu$.

If $F_{X_\alpha}^{X_\beta}$ has a marriage, then, by Theorem 4.2, the family $F_{X_\alpha}^{X_\beta}$ is ρ-positive for each cardinal $\rho < \mu$. By the inductive hypothesis, $F_{X_\alpha}^{X_\beta}$ is ρ-unblocked for each $\rho < \mu$.

To prove the converse, we first observe that, by the inductive hypothesis, $F_{X_\alpha}^{X_\beta}$ is ρ-positive for each $\rho < \mu$. By Theorem 4.4, it is enough to prove $\|F_{X_\alpha}^{X_\beta}\| < \mu$. Let A be a set and K be critical in $F \setminus A$

such that $|A| < \mu$ and $X_\beta = A \cup F(K)$. By Lemma 2.5, there is a set $L \subseteq K$ critical in $F \setminus (A \cup X_\alpha)$ such that

(*) $\quad F(L) \cup A \cup X_\alpha = F(K) \cup A \cup X_\alpha = X_\beta$.

If $i \in L$, then $F_{X_\alpha}(i) \neq \emptyset$. Thus L is critical in $F_{X_\alpha} \setminus (A \setminus X_\alpha)$. By property (*), $F(L) \subseteq X_\beta$. Therefore L is critical in $F_{X_\alpha}^{X_\beta} \setminus (A \setminus X_\alpha)$. Furthermore, if $i \in \operatorname{dom} F_{X_\alpha}^{X_\beta}$, then $F_{X_\alpha}^{X_\beta}(i) \subseteq X_\beta \setminus X_\alpha = F_{X_\alpha}(L) \cup (A \setminus X_\alpha) =$ $= F_{X_\alpha}^{X_\beta}(L) \cup (A \setminus X_\alpha)$. It follows that $F_{X_\alpha}^{X_\beta}(\operatorname{dom} F_{X_\alpha}^{X_\beta}) \subseteq F_{X_\alpha}^{X_\beta}(L) \cup (A \setminus X_\alpha)$ and so $\| F_{X_\alpha}^{X_\beta} \| \leq |A \setminus X_\alpha| < \mu$.

§5. The criterion of Aharoni, Nash-Williams, and Shelah

The first necessary and sufficient condition for the existence of an injective choice function for an arbitrary family $(F(i) : i \in I)$ of sets was proved by Aharoni, Nash-Williams, and Shelah [ANS1]. This criterion and especially Shelah's paper [S2] were the yardstick for our considerations in Section 4, and so their criteria are similar to our μ-test criterion. Both criteria forbid certain substructures for the existence of an injective choice function. The aim of this section is the proof that the existence of an obstruction or of an impediment in the sense of [ANS1] implies the existence of a negative μ-test or of sets K and A such that K is critical in F_A and $|\{i \in I \setminus K : F(i) \subseteq F(K) \cup A\}| > |A|$ (see p. 48) and vice versa. The notion of obstruction is more elegant and easier to handle, but for technical reasons we also introduce the equivalent notion of impediment.

First we need some further definitions for bipartite graphs.

Definition. Let $\Gamma = (M,W,E)$ be a bipartite graph. If $\Gamma_1 = (M_1,W_1,E_1)$ and $\Gamma_2 = (M_2,W_2,E_2)$ are subgraphs of Γ, then we define the union, intersection, difference, and join of Γ_1 and Γ_2 to be respectively $\Gamma_1 \cup \Gamma_2 = (M_1 \cup M_2, W_1 \cup W_2, E_1 \cup E_2)$, $\Gamma_1 \cap \Gamma_2 = (M_1 \cap M_2, W_1 \cap W_2, E_1 \cap E_2)$, $\Gamma_1 \setminus \Gamma_2 = \Gamma_1(M_1 \setminus M_2, W_1 \setminus W_2)$ and $\Gamma_1 \vee \Gamma_2 = \Gamma(M_1 \cup M_2, W_1 \cup W_2)$. Γ_1 and Γ_2 are called disjoint if $\Gamma_1 \cap \Gamma_2 = (\emptyset,\emptyset,\emptyset)$. Similarly, if $\overline{\Sigma} = (\Sigma_i : i \in I)$ is any family of subgraphs of Γ, then $\cup\overline{\Sigma} = \cup\{\Sigma_i : i \in I\}$, $\cap\overline{\Sigma} = \cap\{\Sigma_i : i \in I\}$ have their obvious meaning and $\vee\overline{\Sigma} = \vee\{\Sigma_i : i \in I\} =$ $= \Gamma(\cup\{M_{\Sigma_i} : i \in I\}, \cup\{W_{\Sigma_i} : i \in I\})$. A sequence of subgraphs of Γ will usually be denoted by one of the Greek capitals Γ, Λ, Σ or Π with a bar on top. If $\overline{\Pi}$ is such a sequence, then we always write Π_α to denote the α-th term of this sequence.

A sequence $(\Pi_\alpha : \alpha \leq \xi)$ of saturated subgraphs of Γ is called a <u>ξ-tower</u> (in Γ) if $\Pi_\alpha \leq \Pi_\beta$ for $\alpha < \beta \leq \xi$ and $\Pi_\alpha = \cup\{\Pi_\beta : \beta < \alpha\}$ for every limit ordinal $\alpha \leq \xi$.

A <u>ξ-ladder</u> in Γ is a sequence $\overline{\Lambda} = (\Lambda_\alpha : \alpha < \xi)$ of pairwise disjoint subgraphs of Γ such that $\overline{\Pi} = (\Pi_\alpha : \alpha \leq \xi)$ is a ξ-tower where $\Pi_\alpha = \vee\{\Lambda_\beta : \beta < \alpha\}$. $\overline{\Pi}$ is called <u>the tower associated with $\overline{\Lambda}$</u>. We call every Λ_α a <u>rung</u> of the ladder $\overline{\Lambda}$. If $\overline{\Pi} = (\Pi_\alpha : \alpha \leq \xi)$ is a ξ-tower, then $\overline{\Lambda} = (\Lambda_\alpha : \alpha < \xi)$ is the <u>associated ξ-ladder</u> where $\Lambda_\alpha = \Pi_{\alpha+1} \setminus \Pi_\alpha$ $(\alpha < \xi)$. $\overline{\Pi}$ is a tower (ladder) if it is a ξ-tower (ξ-ladder) for some ordinal ξ.

<u>Remark</u>. If you imagine a "real" ladder, then it is better to realize Λ_α as the space between two rungs. Since the associated tower consists of saturated subgraphs, we can state a trivial but very important fact: There are no edges from any M_{Λ_α} to a rung of higher level. Any edge in Γ starting from M_{Λ_α} must lead to a W_{Λ_β} with $\beta \leq \alpha$. This intuitive concept shall yield in practice applications of Fodor's Theorem.

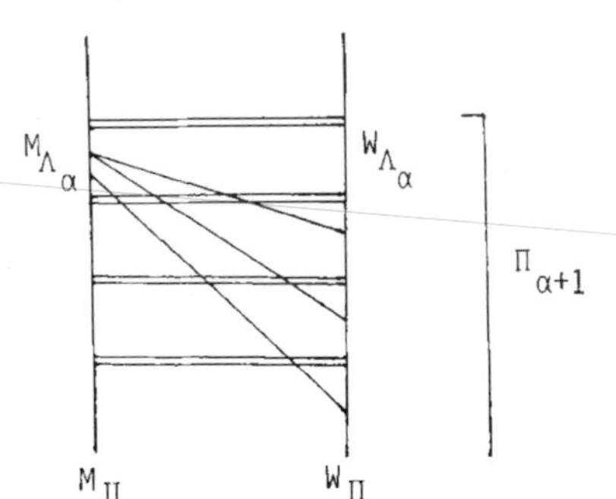

<u>Definition</u>.
Let $Rc = \{1\} \cup \{\varkappa : \varkappa > \aleph_0$ and $\varkappa$ is a regular cardinal$\}$. We define, by transfinite induction on $\varkappa \in Rc$, the property of a bipartite graph Π of being a $\varkappa$-impediment ($\varkappa$-obstruction) . We call Π a <u>1-impediment</u> or <u>1-obstruction</u> if $\Pi \setminus \{a\}$ is critical for some $a \in M_\Pi$. If $\varkappa \in Rc \setminus \{1\}$, then we say that Π is a <u>$\varkappa$-impediment</u> if there is a $\varkappa$-ladder $\overline{\Lambda} = (\Lambda_\alpha : \alpha < \varkappa)$ in Π with $\Pi = \vee\{\Lambda_\alpha : \alpha < \varkappa\}$ which has the following properties:

(I1) For each $\alpha < \varkappa$, the rung Λ_α of $\overline{\Lambda}$ is either
  (i) a μ-impediment for some $\mu \in Rc \cap \varkappa$ or
  (ii) of the form $(\emptyset, \{w\}, \emptyset)$ for some $w \in W_\Pi$ or
  (iii) critical.

(I2) The set $S = \{\alpha < \varkappa : \Lambda_\alpha$ is a μ-impediment for some $\mu < \varkappa\}$ is stationary in $\varkappa$.

If $\varkappa \in Rc$ and $\varkappa \neq 1$, then we call Π a $\varkappa$-obstruction if (I1) and (I2) hold if you delete (iii). We call the corresponding properties (O1) and (O2). $\overline{\Lambda}$ is called an impeding ladder (obstructive ladder) for Π. Π is said to be an impediment (obstruction) if there exists a $\varkappa \in Rc$ such that Π is a $\varkappa$-impediment (a $\varkappa$-obstruction). We call Π an impediment (obstruction) in the bipartite graph Γ if Π is a saturated subgraph of Γ and Π is an impediment (obstruction).

Now we can state the criterion of Aharoni, Nash-Williams and Shelah:

Theorem 5.1. A bipartite graph Γ has a marriage if and only if there exists no impediment in Γ.

To give a first impression, we discuss briefly the case of an $\aleph_1$-impediment. Let Π be an $\aleph_1$-impediment, and let $\overline{\Lambda}$, S be the objects which guarantee (I1) and (I2). Take $\Lambda_\alpha = (M_\alpha, W_\alpha, E_\alpha)$, let $\overline{\Pi}$ be the tower associated with $\overline{\Lambda}$ and assume, to get a contradiction, that there is a marriage f of Π. For any $\alpha \in S$, the subgraph Λ_α has no marriage and thus there is an $a_\alpha \in M_\alpha$ such that $f(a_\alpha) \notin W_\alpha$. Now $\Pi_{\alpha+1}$ is saturated, so $f(a_\alpha) \in W_{\varphi(\alpha)}$ for some ordinal $\varphi(\alpha) < \alpha$. Since S is stationary and $\varphi : S \to \varkappa$ is regressive, there exist, by Fodor's Theorem, a stationary subset S^* of S and a $\gamma < \varkappa$ such that $f(a_\alpha) \in W_\gamma$ for all $\alpha \in S^*$. Hence Λ_γ cannot have the form $(\emptyset, \{w\}, \emptyset)$, so $\Lambda_\gamma \setminus N$ must be critical for some $N \subseteq M_\gamma$ with $|N| \leq 1$ by property (I1). $\aleph_1$ women of W_γ are married with the men a_α, $\alpha \in S^*$, therefore $\aleph_1$ men of M_γ are forced to marry a woman in $\bigcup\{W_\beta : \beta < \gamma\}$: If one puts $A = \emptyset$, $K = M_\gamma \setminus N$, $B = \{a_\alpha : \alpha \in S^*\}$, and $J = \{m \in K : f(m) \in W_\gamma\}$, then this fact follows by Theorem 2.10. Since $|\gamma| < \aleph_1$ and $\aleph_1$ is regular, there exists a $\gamma_1 < \gamma$ such that $\aleph_1$ men of M_γ marry women in W_{γ_1}. The same argument yields a $\gamma_2 < \gamma_1$ such that $\aleph_1$ men of M_{γ_1} marry women in W_{γ_2} and so on. So we get an infinite strictly descending sequence of ordinals, and this is a contradiction.

For the proof of Theorem 5.1 we need first two lemmas from [ANS1].

Lemma 5.2. If Σ is a saturated critical subgraph of Γ and if Π is a $\varkappa$-obstruction in $\Gamma \setminus \Sigma$, then $\Pi \vee \Sigma$ is a $\varkappa$-obstruction in Γ.

Proof. We prove the lemma by transfinite induction on $\varkappa \in Rc$. First let $\varkappa = 1$. Since $\Pi \setminus \{a\}$ is critical for some $a \in M_{\Pi}$, we have that $\Sigma \vee (\Pi \setminus \{a\}) = (\Sigma \vee \Pi) \setminus \{a\}$ is critical, too. So $\Sigma \vee \Pi$ is a 1-obstruction.

Now let $\varkappa \in Rc \setminus \{1\}$ and let $\bar{\Lambda}$ be a $\varkappa$-ladder in $\Gamma \setminus \Sigma$ with the corresponding tower $\bar{\Pi}$ and the stationary set S, which guarantee properties (O1) and (O2). Take $\xi = \min S$ and define $\Lambda'_\alpha = \Lambda_\alpha$, if $\alpha \neq \xi$, $\Lambda'_\xi = \Lambda_\xi \vee \Sigma$. By the induction hypothesis, Λ'_ξ is a μ-obstruction, since $\xi \in S$ and Λ_ξ is a μ-obstruction for some $\mu < \varkappa$. Further $\Pi'_\alpha = \vee \{\Lambda'_\beta : \beta < \alpha\}$ is saturated in $\Pi \vee \Sigma$, since Π_α is saturated in Π, $\Pi = \vee \bar{\Pi}$, and Σ and Π are disjoint. Consequently $(\Lambda'_\alpha : \alpha < \varkappa)$ is a $\varkappa$-ladder for $\Pi \vee \Sigma$ which yields, with the same S, the properties (O1) and (O2).

Lemma 5.3. Π is a $\varkappa$-obstruction if and only if Π is a $\varkappa$-impediment.

Proof. Obviously we only need to show that every $\varkappa$-impediment Π is a $\varkappa$-obstruction. We prove this by induction on $\varkappa$. For $\varkappa = 1$ there is nothing to show. Let $\varkappa \in Rc \setminus \{1\}$, let $\bar{\Lambda} = (\Lambda_\alpha : \alpha < \varkappa)$ be an impeding $\varkappa$-ladder for $\Pi = \vee \{\Lambda_\alpha : \alpha < \varkappa\}$, and let S_1, S_2, and S_3 be respectively the set of those $\alpha < \varkappa$ such that Λ_α has the property (i), (ii), and (iii) in (I1). For successive elements $\gamma_1, \gamma_2 \in S_1$ with $\gamma_1 < \gamma_2$ we collect those Λ_α for which $\alpha \in S_3$ and $\gamma_1 < \alpha < \gamma_2$, and form a new rung $\Delta_{f(\gamma_2)}$, where f is the well ordering isomorphism from $S_1 \cup S_2$ onto $\varkappa$. For $\alpha \in S_2$ we take $\Delta_{f(\alpha)} = \Lambda_\alpha$. The ladder $\bar{\Delta} = (\Delta_\alpha : \alpha < \varkappa)$ will have the desired properties. In detail: Let $g : S_3 \to S_1$ be defined by $g(\alpha) = \min(S_1 \setminus \alpha)$. For $\alpha \in S_1$ take $\Lambda^*_\alpha = \vee \{\Lambda_\beta : \alpha = g(\beta)\}$ and $\Delta_{f(\alpha)} = \Lambda_\alpha \vee \Lambda^*_\alpha$, for $\alpha \in S_2$ take $\Delta_{f(\alpha)} = \Lambda_\alpha$. Obviously Λ^*_α is critical for every $\alpha \in S_1$. We show that $\bar{\Delta} = (\Delta_\alpha : \alpha < \varkappa)$ is a $\varkappa$-ladder with the associated tower $\bar{\Sigma} = (\Sigma_\alpha : \alpha \leq \varkappa)$, $\Sigma_\alpha = \vee \{\Delta_\beta : \beta < \alpha\}$.

Claim 1: If $\bar{\Pi} = (\Pi_\alpha : \alpha \leq \varkappa)$ is the tower associated with $\bar{\Lambda}$ and if $\alpha \in S_1$, then $\Pi_\alpha = \Sigma_{f(\alpha)} \vee \Lambda^*_\alpha$ and $\Pi_{\alpha+1} = \Sigma_{f(\alpha)+1}$.

Clearly we have

$$\Sigma_{f(\alpha)} = \vee \{\Delta_{f(\beta)} : \beta \in S_1 \cup \alpha\} \vee \vee \{\Delta_{f(\beta)} : \beta \in S_2 \cap \alpha\}.$$

By the definition of $\Delta_{f(\beta)}$, we get

$$\Sigma_{f(\alpha)} = \vee\{\Lambda_\beta : \beta \in S_1 \cap \alpha\} \vee \vee\{\Lambda^*_\beta : \beta \in S_1 \cap \alpha\} \vee \vee\{\Lambda_\beta : \beta \in S_2 \cap \alpha\}.$$

By the definition of Λ^*_β, we have

$$\vee\{\Lambda^*_\beta : \beta \leq \alpha, \beta \in S_1\} = \vee\{\Lambda_\beta : \beta \in S_3 \cap \alpha\}.$$

Consequently

$$\Sigma_{f(\alpha)} \vee \Lambda^*_\alpha = \vee\{\Lambda_\beta : \beta < \alpha, \beta \in S_1 \cup S_2 \cup S_3\} = \Pi_\alpha$$

and

$$\Sigma_{f(\alpha)+1} = \Sigma_{f(\alpha)} \vee \Delta_{f(\alpha)} = \Sigma_{f(\alpha)} \vee \Lambda^*_\alpha \vee \Lambda_\alpha = \Pi_\alpha \vee \Lambda_\alpha = \Pi_{\alpha+1},$$

and Claim 1 is proved. Claim 1 implies at once that each $\Sigma_{\gamma+1}$ is saturated, since $\bar{\Pi}$ is a sequence of saturated subgraphs of Π. With this fact we see by transfinite induction that $\bar{\Sigma}$ is a $\varkappa$-tower. Further we have $\vee\{\Delta_\beta : \beta < \varkappa\} = \vee\bar{\Sigma} = \vee\bar{\Pi} = \Pi$. If $\alpha \in S_2$, then $\Delta_{f(\alpha)}$ has the form $(\emptyset, \{w\}, \emptyset)$. Our proof is completed if we can show that for every $\alpha \in S_1$ the rung $\Delta_{f(\alpha)}$ is a ρ-obstruction for some $\rho < \varkappa$, and that $f[S_1]$ is stationary in $\varkappa$. Now we have that S_1 is stationary by (I2). By Lemma I.6.4b, the set $\{\alpha < \varkappa : f(\alpha) = \alpha\} = \{\alpha < \varkappa : f^{-1}(\alpha) = \alpha\}$ is a club in $\varkappa$, and so $S_1 \cap \{\alpha < \varkappa : f(\alpha) = \alpha\}$ is a stationary subset of $f[S_1]$, hence $f[S_1]$ is stationary.

Claim 2: If $\alpha \in S_1$ and Λ_α is a ρ-impediment for some $\rho < \varkappa$, then $\Delta_{f(\alpha)}$ is a ρ-obstruction.

By the induction hypothesis, Λ_α is a ρ-obstruction, and in addition Λ^*_α is critical. Further we have that Λ_α and Λ^*_α are disjoint, since every β with $g(\beta) = \alpha$ is smaller than α. So Claim 2 follows from Lemma 5.2.

Lemma 5.4.

If Π is a $\varkappa$-impediment ($\varkappa$-obstruction), then there are sets $J \subseteq M_\Pi$ and $A \subseteq W_\Pi$ such that

(i) $|J| = \varkappa$ and $|J \cup A| = \varkappa$ and

(ii) $(\Pi \setminus J) \setminus A$ is critical .

Proof. If $\varkappa = 1$, then the definition of an impediment yields J with $|J| = 1$ and $A = \emptyset$. Let $\varkappa \in Rc \setminus \{1\}$ and assume that our assertion is true for every $\lambda \in Rc \cap \varkappa$. Let $\bar{\Lambda} = (\Lambda_\alpha : \alpha < \varkappa)$ be an impeding $\varkappa$-ladder for Π, $\Pi = \vee\bar{\Lambda}$, such that (I1) and (I2) hold. If Λ_α is a μ_α-impediment for some $\mu_\alpha < \varkappa$, then there exist, by the inductive hypothesis, sets $J_\alpha \subseteq M_{\Lambda_\alpha}$ and $A_\alpha \subseteq W_{\Lambda_\alpha}$ such that $|A_\alpha \cup J_\alpha| = \mu_\alpha$,

$|J_\alpha| = \mu_\alpha$ and $(\Lambda_\alpha \setminus J_\alpha)\setminus A_\alpha$ is critical. If $\Lambda_\alpha = (\emptyset,\{w\},\emptyset)$, then let $J_\alpha = \emptyset$ and $A_\alpha = \{w\}$. If Λ_α is critical , then take $A_\alpha = J_\alpha = \emptyset$. Now define $J = \cup\{J_\alpha : \alpha < \varkappa\}$ and $A = \cup\{A_\alpha : \alpha < \varkappa\}$. Property (I2) implies that $|J| = \varkappa$ and $|J \cup A| = \varkappa$. Further $(\Pi \setminus J)\setminus A = \vee\{(\Lambda_\alpha \setminus J_\alpha)\setminus A_\alpha : \alpha < \varkappa\}$ is critical as a disjoint join of critical graphs. The proof for $\varkappa$-obstructions is established completely analogously.

Lemma 5.5. Let $\Pi \leq \Gamma$ and $\Pi \leq \Delta$, and let $E_\Gamma\langle m\rangle = E_\Delta\langle m\rangle$ for every $m \in M_\Pi$. Then Π is a $\varkappa$-impediment ($\varkappa$-obstruction) in Γ if and only if Π is a $\varkappa$-impediment ($\varkappa$-obstruction) in Δ.

Proof. The hypotheses guarantee that Π is saturated in Γ if and only if Π is saturated in Δ. The assertion now follows easily by transfinite induction on $\varkappa$.

We want to show that a family $F = (F(i) : i \in I)$ is λ-negative, i.e. not λ-positive for some λ if and only if the corresponding graph Γ_F contains an impediment. With Theorem 4.2 and Theorem 4.4 this yields at once Theorem 5.1. First we reformulate the property that F is λ-unblocked which is equivalent to the property that F is λ-positive by Theorem 4.14. Of course F is called λ-blocked if F is not λ-unblocked.

Lemma 5.6.
A family $F = (F(i) : i \in I)$ is λ-blocked if and only if one of the following four properties holds for F:

(1) There exist a set A with $|A| \leq \lambda$ and a set K which is critical in F_A such that
$$|\{i \in I \setminus K : F(i) \subseteq F(K) \cup A\}| > |A|.$$

(2) There exist an uncountable regular cardinal $\mu \leq \lambda$ and a μ-test $(X_\alpha : \alpha < \mu)$ such that the set
$$S = \{\alpha < \mu : \exists i (F(i) \subseteq X_\alpha \wedge \forall \beta < \alpha (F(i) \not\subseteq X_\beta))\}$$
is stationary in μ.

(3) There exist an uncountable regular cardinal $\mu \leq \lambda$ and a μ-test $(X_\alpha : \alpha < \mu)$ such that the set
$$S = \{\alpha < \mu : F_{X_\alpha}^{X_\beta} \text{ is } \rho\text{-blocked for some } \rho < \mu \text{ and some } \beta \geq \alpha\}$$
is stationary in μ.

(4) There exist an uncountable regular cardinal $\mu \leq \lambda$, a set B with $|B| < \mu$, and a set L which is critical in F_B such that $F^{F(L) \cup B}$ is ρ-blocked for some $\rho < \mu$.

Proof. This is an immediate consequence of the definition of the property of being λ-unblocked if you keep in mind that one of the sets S_1 and S_2 must be stationary, if $S_1 \cup S_2$ is stationary, and if you observe that the negation of "F^{X_β} is ρ-unblocked for every $\rho<\mu$ and every $\beta<\mu$" is independent of a μ-test, as we formulated it in (4).

From now on up to the end of the proof we speak of (j) instead of "property (j) in Lemma 5.6", $j \in \{1,\dots,4\}$.

Theorem 5.7. Let $F = (F(i) : i \in I)$ be a family, and let Γ_F be the corresponding bipartite graph. Then F is λ-blocked for some cardinal λ if and only if Γ_F contains an impediment.

The proof of Theorem 5.7 is rather technical though its ideas are not so complicated. One direction is proved in a - let us say - natural way: If Π is a $\varkappa$-impediment in Γ_F, then the critical subgraphs yielded by Lemma 5.4 lead to a $\varkappa$-test. And if properties (2), (3), (4) are false, Fodor's Theorem guarantees the existence of sets A and K which satisfy (1). The details can be found in Lemma 5.10.

The proof of the other direction of Theorem 5.7 will be led by transfinite induction on λ. The easy case is (1), and there we need no induction - see Lemma 5.8. In the inductive proof we assume that F has not property (1). This will guarantee that the impediments of the families in (3) and (4) to which the hypothesis of the induction is applied do not become too large - a λ-blocked family may contain, by property (1), a λ^+-impediment, where λ^+ is the cardinal successor of λ. This is done in Lemma 5.9. For the rest of the proof we need the following

Definition. If $F = (F(i) : i \in I)$ is a family and if there exist sets A and K such that K is critical in F_A and $|\{i \in I \setminus K : F(i) \subseteq F(K) \cup A\}| > |A|$, then let b(F) be the least cardinal $|A|$ such that there is a set K critical in F_A such that $|\{i \in I \setminus K : F(i) \subseteq F(K) \cup A\}| > |A|$. Otherwise let $b(F) = -1$.

For example $b(F) = 0$ if there exists an $i \in I$ such that $F(i) = \emptyset$ - choose $A = \emptyset$ and $K = \emptyset$.

Lemma 5.8. Let $F = (F(i) : i \in I)$ be a family such that $b(F) \geq 0$. Then Γ_F possesses a $b(F)^+$-obstruction, if $b(F)$ is infinite, and a 1-obstruction, if $b(F)$ is finite.

Proof. Choose A and K according to the definition of $b(F)$ and let $\nu = b(F)$, $\Gamma = \Gamma_F$.

Case 1. $\nu \geq \aleph_0$.

Let $(a_\alpha : \alpha < \nu)$ be an enumeration of A and $\Lambda_\alpha = (\emptyset, \{a_\alpha\}, \emptyset)$ for $\alpha < \nu$. Now choose $i_0 \in I \setminus K$ such that $F(i_0) \subseteq F(K) \cup A$ and define $\Lambda_\nu = \Gamma(K \cup \{i_0\}, F(K) \setminus A)$. Since K is critical in F_A, Λ_ν is a 1-obstruction. By the choice of A and K, there exists an injective sequence $(i_\alpha : \nu < \alpha < \nu^+)$ of elements of $I \setminus (K \cup \{i_0\})$ such that $F(i_\alpha) \subseteq F(K) \cup A$. Let $\Lambda_\alpha = (\{i_\alpha\}, \emptyset, \emptyset)$ for $\nu < \alpha < \nu^+$ and $\overline{\Lambda} = (\Lambda_\alpha : \alpha < \nu^+)$. $\overline{\Lambda}$ is a ν^+-ladder, for it is easy to see that the subgraphs $\Pi_\alpha = \bigvee\{\Lambda_\beta : \beta < \alpha\}$ are saturated in Γ. For any α with $\alpha \geq \nu$, Λ_α is a 1-obstruction, especially the set of these α is stationary, and so $\overline{\Lambda}$ is an obstructive ladder for $\Pi = \bigvee\{\Lambda_\beta : \beta < \nu^+\}$: Π is a ν^+-obstruction.

Case 2. $\nu < \aleph_0$.

Choose $J \subseteq \{i \in I \setminus K : F(i) \subseteq F(K) \cup A\}$ such that $|J| = |A| + 1 = n+1$. By Lemma 4.1, the family $F \restriction K \cup J$ has no marriage. If $J = \{i_1, \ldots, i_{n+1}\}$, then let k be the least natural number l such that $F \restriction K \cup \{i_1, \ldots, i_l\}$ has no marriage. Then $F(i_k) \subseteq \operatorname{rng} h$ for every marriage h of $F \restriction K \cup \{i_1, \ldots, i_{k-1}\}$, and especially $F(i_k) \subseteq F(K \cup \{i_1, \ldots, i_{k-1}\})$. By Corollary 2.13, there exists a set $L \subseteq K \cup \{i_1, \ldots, i_{k-1}\}$ such that L is critical in F and $F(i_k) \subseteq F(L)$. Then $\Gamma(L \cup \{i_k\}, F(L))$ is a 1-obstruction.

Lemma 5.9.

Let $F = (F(i) : i \in I)$ be λ-blocked such that F does not have property (1) of Lemma 5.6, i.e. $b(F) = -1$. If F has property (2) or (3) respectively, let $(X_\alpha : \alpha < \mu)$ be the corresponding μ-test. Then the following holds:

(i) If $\beta > \alpha$ and $G = F_{X_\alpha}^{X_\beta}$, then $b(G)^+ < \mu$.

(ii) Γ_F contains a $\varkappa$-impediment for some $\varkappa \leq \mu$.

Proof. For (i), let us assume, to get a contradiction, that $\nu^+ \geq \mu$, $\nu = b(G)$. According to the definition of a μ-test, there exist sets A and K such that $|A| < \mu$, K is critical in F_A and $X_\alpha = F(K) \cup A$.

Since $b(G) > -1$, we can choose sets B and K^* such that K^* is critical in G_B and $|B|^+ \geq \mu$ and $|\{i \in \operatorname{dom} G \setminus K^* : G(i) \subseteq G(K^*) \cup B\}| > |B|$. Define $J_1 = \{i \in \operatorname{dom} G \setminus K^* : G(i) \subseteq G(K^*) \cup B\}$; so we have

$$(*) \quad |J_1| > |B|.$$

Now $G_B = (F^{X_\beta}_{X_\alpha})_B = ((F^{X_\beta})_{X_\alpha})_B = (F^{X_\beta})_{X_\alpha \cup B}$, since $X_\alpha \subseteq X_\beta$. Because of $F(K) \cup A = F_A(K) \cup A$, the set K^* is critical in $F_{A \cup B \cup F_A(K)}$. K is critical in F_A, consequently there exists, by Lemma 2.5, a set $L \subseteq K$ such that L is critical in $(F_A)_B = F_{A \cup B}$ and $F_A(L) \cup A \cup B = F_A(K) \cup A \cup B$. So K^* is critical in $F_{A \cup B \cup F_A(L)} = F_{A \cup B \cup F_{A \cup B}(L)}$. Now we apply Lemma 2.6 to $F_{A \cup B}$, L, and K^* and get that $L \cup K^*$ is critical in $F_{A \cup B}$. Define $J_2 = \{i \in I \setminus (L \cup K^*) : F(i) \subseteq F(L \cup K^*) \cup A \cup B\}$. If we succeed in proving $J_1 \subseteq J_2$, then we have $|J_2| > |B|$ by $(*)$. Further $|B|^+ \geq \mu$ and $|A| < \mu$, so $|B \cup A| = |B|$ and $|J_2| > |A \cup B|$. Consequently $L \cup K^*$ is critical in $F_{A \cup B}$ and $|J_2| > |A \cup B|$, which contradicts the hypothesis that $b(F) = -1$.

There remains the proof of $J_1 \subseteq J_2$. Let $i \in J_1$. Of course $i \notin K^*$. If $i \in L$, then $i \in K$ and so $F(i) \subseteq X_\alpha$ which contradicts $i \in \operatorname{dom} G$. Consequently $i \in I \setminus (L \cup K^*)$. Further $G(i) \subseteq G(K^*) \cup B$ implies that $F(i) \subseteq (F(K^*) \setminus X_\alpha) \cup X_\alpha \cup B \subseteq A \cup B \cup F(K^*) \cup F(K) = A \cup B \cup F(K^* \cup L)$, since $F(K) \cup A \cup B = F(L) \cup A \cup B$. So we obtain $i \in J_2$, and (i) is proved.

Now we come to the proof of (ii).
We prove (ii) be transfinite induction on λ and assume as induction hypothesis that the assertion is true for any ρ-blocked family H with $b(H) = -1$ and $\rho < \lambda$. Notice that $b(F) = -1$ implies that λ must be uncountable.

Case 1. F has property (4).

Let $\mu \geq \aleph_1$ be regular, $|B| < \mu$, and L be critical in F_B such that $H = F^{F(L) \cup B}$ is ρ-blocked for some $\rho < \mu$. Since $b(F) = -1$ and H is a subfamily of F, we have $b(H) = -1$. So Γ_H contains an $\varkappa$-impediment Π for some $\varkappa \leq \rho$ by the induction hypothesis. Lemma 5.5 shows that Π is an $\varkappa$-impediment in Γ_F.

Case 2. F has property (2).

By the definition of a μ-test, there exist sets A_α and K_α such that K_α is critical in F_{A_α}, $|A_\alpha| < \mu$, and $X_\alpha = F(K_\alpha) \cup A_\alpha$ for each $\alpha < \mu$. Since $S = \{\alpha < \mu : \exists i\, (F(i) \subseteq X_\alpha \wedge \forall \beta < \alpha\, (F(i) \not\subseteq X_\beta))\}$ is stationary, the same holds for $S^* = S \cap \{\alpha : \alpha$ is a limit ordinal$\}$. For each $\alpha \in S^*$ choose an $i_\alpha \in I$ such that $F(i_\alpha) \subseteq X_\alpha$ and $F(i_\alpha) \not\subseteq X_\beta$ for every $\beta < \alpha$. Clearly $i_\alpha \neq i_\beta$ for $\alpha \neq \beta$. We want to construct an obstructive ladder $\overline{\Lambda}$, and it suggests itself to choose $\Lambda_\alpha = (\{i_\alpha\}, \emptyset, \emptyset)$ for $\alpha \in S^*$. First these are 1-obstructions, second S^* is stationary, and third there is some hope of getting a tower of saturated subgraphs if we succeed in defining the rungs Λ_β, $\beta \notin S^*$, in such a way that $X_\alpha = \cup\{W_{\Lambda_\beta} : \beta < \alpha\}$ for any $\alpha \in S^*$, since $F(i_\alpha) \subseteq X_\alpha$. Now $X_\beta = F(K_\beta) \cup A_\beta$, where K_β is critical in F_{A_β}. Roughly speaking, the elements of $A_{\beta+1}$ X_β are yielded by rungs of the form $(\emptyset, \{w\}, \emptyset)$, and $X_{\beta+1} \setminus (A_{\beta+1} \cup X_\beta)$ is the set of women of a critical rung. Since $(X_\alpha : \alpha < \mu)$ is continuous, this also yields the elements of $X_\alpha = \cup\{X_{\beta+1} : \beta < \alpha\}$ for limit ordinals α. The problem that the rungs must be disjoint is solved by Lemma 2.5, which says that if you remove a set of women Z from a critical subfamily G of F, then the remaining women of G are in the range of a critical subfamily of $F \setminus Z$, and this is just the fact we need. Let us come to the technical details: For each $\beta < \mu$ there exists, by Lemma 2.5, a subset $L_{\beta+1} \subseteq K_{\beta+1}$ which is critical in $(F_{A_{\beta+1}})_{X_\beta \cup A_{\beta+1}} = F_{X_\beta \cup A_{\beta+1}}$ and which has the property $F(L_{\beta+1}) \cup X_\beta \cup A_{\beta+1} = F(K_{\beta+1}) \cup X_\beta \cup A_{\beta+1} = X_{\beta+1} \cup X_\beta = X_{\beta+1}$. Further let $(a_\nu^\beta : \nu < \gamma_\beta)$ be an enumeration of $A_{\beta+1} \setminus X_\beta$ and $B_\nu^\beta = (\emptyset, \{a_\nu^\beta\}, \emptyset)$ for every $\beta < \mu$ and $\nu < \gamma_\beta$. Now define, for each $\alpha < \mu$, a sequence C_α of subgraphs of Γ. If $\alpha \in S^*$, then $C_\alpha = ((\{i_\alpha\}, \emptyset, \emptyset))$ is a sequence of length 1. If α is limit ordinal not in S^* or if $\alpha = 0$, then $C_\alpha = ((\emptyset, \emptyset, \emptyset))$. If $\alpha = \beta + 1$ is a successor ordinal then define

$$C_\alpha = (B_\nu^\beta : \nu < \gamma_\beta) * (\Gamma(L_{\beta+1}, F(L_{\beta+1}) \setminus (X_\beta \cup A_{\beta+1}))).$$

Let $\overline{\Lambda}$ be the sequence of length μ which results from the concatenation of the sequences C_α, $\alpha < \mu$, take $\overline{\Lambda} = (\Lambda_\beta : \beta < \mu)$, and let $\overline{\Pi} = (\Pi_\alpha : \alpha \leq \mu)$ be the corresponding tower. The construction shows that the rungs of $\overline{\Lambda}$ are pairwise disjoint and that the subgraphs $\Pi_\alpha = \vee\{\Lambda_\beta : \beta < \alpha\}$ are saturated in Γ. Therefore $\overline{\Lambda}$ is a ladder whose rungs have the form (i), (ii), or (iii) of property (I1) in the definition of an impediment. It remains to show that (I2) is fulfilled. Define $\varphi : \mu \to \mu$ inductively by $\varphi(0) = 0$ and $\varphi(\alpha) = \sup\{\varphi(\beta) : \beta < \alpha\}$, if α is a limit ordinal; if $\alpha = \beta + 1$ is a successor ordinal, then $\varphi(\alpha) = \varphi(\beta) + 1$ if β is a limit ordinal (we add the sequence C_α of length 1), and otherwise $\varphi(\alpha) = \varphi(\beta) + \gamma_\beta + 1$ as the ordinal sum of $\varphi(\beta)$ and $\gamma_\beta + 1$ (we add the sequence C_α of length $\gamma_\beta + 1$). Especially $(\{i_\alpha\}, \emptyset, \emptyset) = \Lambda_{\varphi(\alpha)}$ for each $\alpha \in S^*$. Now the function φ is continuous by definition, and it is strictly increasing. Consequently the set C of its fixed points is a club in μ by Lemma I.6.4, and $S^{**} = S^* \cap C$ is stationary in μ. This proves property (I2), so $\Pi = \vee\{\Lambda_\beta : \beta < \mu\}$ is an impediment in Γ and the proof of Case 2 is completed.

Case 3. F has property (3).

Let $(X_\alpha : \alpha<\mu)$ be the μ-test for which the set $S = \{\alpha<\mu : F_{X_\alpha}^{X_\beta}$ is ρ-blocked for some $\rho<\lambda$ and some $\beta\geq\alpha\}$ is stationary in μ. We handle this case similar to Case 2; the difference will be that we need the hypothesis of the induction and that it is more troublesome to make the rungs of our ladder disjoint: If $F_{X_\alpha}^{X_\beta}$ and $F_{X_\gamma}^{X_\delta}$ are blocked for some successive elements $\alpha, \gamma\in S$, we do not know very much about β and δ. The first step therefore is the construction of a sequence $(\nu_\alpha : \alpha<\mu)$ of ordinals less than such that $F_{X_{\nu_\alpha}}^{X_{\nu_{\alpha+1}}}$ is ρ-blocked for some $\rho<\mu$. We then replace our μ-test $(X_\alpha : \alpha<\mu)$ by the μ-test $(Y_\alpha : \alpha<\mu)$ with $Y_\alpha = X_{\nu_\alpha}$ for $\alpha<\mu$. Let $\nu_0 = 0$ and $\nu_\alpha = \sup\{\nu_\beta : \beta<\alpha\}$ for limit ordinals α. If $\alpha = \beta+1$ is a successor ordinal and $\nu_\beta = \beta\in S$, then let ν_α be the smallest $\gamma\geq\alpha$ such that $F_{X_\beta}^{X_\gamma}$ is ρ-blocked for some $\rho<\mu$; otherwise let $\nu_\alpha = \nu_\beta+1$. Clearly the sequence $(\nu_\alpha : \alpha<\mu)$ is continuous and strictly increasing, so the set C of its fixed points is a club in μ, and the set $C\cap S$ is stationary in μ. By definition we have

$C\cap S\subseteq\{\alpha<\mu : F_{X_{\nu_\alpha}}^{X_{\nu_{\alpha+1}}}$ is ρ-blocked for some $\rho<\mu\}$.

Consequently we define $Y_\alpha = X_{\nu_\alpha}$ for $\alpha<\mu$. The sequence $(Y_\alpha : \alpha<\mu)$ is a continuous chain, and, as a subsequence of $(X_\alpha : \alpha<\mu)$, it is a μ-test. Further the set

$$S = \{\alpha<\mu : F_{Y_\alpha}^{Y_{\alpha+1}} \text{ is } \rho_\alpha\text{-blocked for some } \rho_\alpha<\mu\}$$

is stationary in μ. Let $\alpha\in S$ and define $G_\alpha = F_{Y_\alpha}^{Y_{\alpha+1}}$. If $b(G_\alpha)\geq 0$, then part (i) of this lemma implies $b(G_\alpha)^+<\mu$. Consequently Γ_{G_α} contains a $\varkappa_\alpha$-impediment Σ_α by Lemma 5.8 for some $\varkappa_\alpha\leq b(G_\alpha)^+$. If $b(G_\alpha) = -1$ and G_α is ρ_α-blocked for some $\rho_\alpha<\mu$, then the induction hypothesis yields a $\varkappa_\alpha$-impediment Σ_α in Γ_{G_α} for some $\varkappa_\alpha\leq\rho_\alpha$. Let $Y_\alpha = A_\alpha\cup F(K_\alpha)$ such that $|A_\alpha|<\mu$ and $\varkappa_\alpha$ is critical in F_{A_α}. As in the proof of Case 2, we construct a μ-impediment Σ for Γ_F. The rungs $(\{i_\alpha\},\emptyset,\emptyset)$ are now replaced by the impediments Σ_α just defined. Let $S^* = S\cap\{\alpha : \alpha$ is a limit ordinal$\}$, $M_\alpha = M_{\Sigma_\alpha}$, $W_\alpha = W_{\Sigma_\alpha}$ and $\Gamma_\alpha = \Gamma_{G_\alpha}$. As in Case 2 there exists, for each $\alpha\in S^*$, by Lemma 2.5 a set $L_{\alpha+1}\subseteq K_{\alpha+1}$ which is critical in $F\setminus(A_{\alpha+1}\cup Y_\alpha\cup W_\alpha)$ and has the property

$F(L_{\alpha+1}) \cup A_{\alpha+1} \cup Y_\alpha \cup W_\alpha = F(K_{\alpha+1}) \cup A_{\alpha+1} \cup Y_\alpha \cup W_\alpha = Y_{\alpha+1} \cup Y_\alpha \cup W_\alpha = Y_{\alpha+1}$, since $Y_\alpha \cup W_\alpha \subseteq Y_{\alpha+1}$. Therefore we have $L_{\alpha+1} \cup M_\alpha \subseteq \operatorname{dom} G_\alpha$. As above let $(a^\alpha_\nu : \nu<\gamma_\alpha)$ be an enumeration of $A_{\alpha+1} \setminus (Y_\alpha \cup W_\alpha)$, $B^\alpha_\nu = (\emptyset, \{a^\alpha_\nu\}, \emptyset)$ for $\nu<\gamma_\alpha$, and C_α be the sequence $C_\alpha = (\Sigma_\alpha) * (B^\alpha_\nu : \nu<\gamma_\alpha) * (\Gamma(L_{\alpha+1}, F(L_{\alpha+1}) \setminus (A_{\alpha+1} \cup Y_\alpha \cup W_\alpha)))$. So C_α is defined for every $\alpha \in S^*$. If $\alpha \notin S^*$ and α is a limit ordinal or $\alpha = 0$, then define $C_\alpha = (\emptyset, \emptyset, \emptyset)$. If $\alpha \notin S^*$ and α is the successor ordinal $\beta+1$, apply Lemma 2.5 again to get a set $L_\alpha \subseteq K_\alpha$ such that L_α is critical in $F \setminus (A_\alpha \cup Y_\beta)$ and $F(K_\alpha) \cup A_\alpha \cup Y_\beta = F(L_\alpha) \cup A_\alpha \cup Y_\beta$. Consequently $Y_\alpha = F(L_\alpha) \cup A_\alpha \cup Y_\beta$ and $L_\alpha \subseteq \operatorname{dom} G_\beta$. If $(a^\beta_\nu : \nu<\gamma_\beta)$ is an enumeration of $A_\alpha \setminus Y_\beta$ and $B^\beta_\nu = (\emptyset, \{a^\beta_\nu\}, \emptyset)$ for $\nu<\gamma_\beta$, then let C_α be the sequence $C_\alpha = C_{\beta+1} = (B^\beta_\nu : \nu<\gamma_\beta) * (\Gamma(L_\alpha, F(L_\alpha) \setminus (A_\alpha \cup Y_\beta)))$. Again let $\overline{\Lambda}$ be the sequence of length μ which results from the concatenation of the sequences C_α, $\alpha<\mu$. A fixed point argument analogous to the proof of Case 2 shows that there exists a stationary subset S^{**} of S^* such that, for each $\alpha \in S^{**}$, we have $\Lambda_\alpha = \Sigma_\alpha$. Now Σ_α is a $\varkappa_\alpha$-impediment in Γ_α for some $\varkappa_\alpha<\mu$, $\Gamma_\alpha \leq \Gamma \setminus Y_\alpha$, and for every $m \in M_\alpha$ clearly $E_{\Gamma\setminus Y_\alpha}\langle m\rangle = E_{\Gamma_\alpha}\langle m\rangle$. Lemma 5.5 shows that Σ_α is a $\varkappa_\alpha$-impediment in $\Gamma \setminus Y_\alpha$, and property (I2) of an impediment is satisfied. The construction of $\overline{\Lambda}$ shows that the subgraphs Λ_β, $\beta<\mu$, are pairwise disjoint. Further every Σ_γ is saturated in Γ_γ and hence in $\Gamma \setminus Y_\gamma$ for every $\gamma \in S^*$, and this implies easily that $\Pi_\alpha = \vee\{\Lambda_\nu : \nu<\alpha\}$ is saturated in Γ for every $\alpha<\mu$. Therefore $\overline{\Lambda}$ is an obstructive ladder for $\Pi = \vee\{\Lambda_\beta : \beta<\mu\}$ and Π is a μ-impediment in Γ. This completes the proof of Lemma 5.9.

<u>Lemma 5.10</u>. If Π is a $\varkappa$-impediment in Γ_F, then the family $F = (F(i) : i \in I)$ is $\varkappa$-blocked.

<u>Proof</u>. We prove the lemma by transfinite induction on $\varkappa$. If $\varkappa = 1$, then there exists an $m \in M_\Pi$ such that $\Pi \setminus \{m\}$ is critical. Since Π is saturated, the sets $A = \emptyset$ and $K = M_\Pi \setminus \{m\}$ yield property (1) of Lemma 5.6. Now let $\varkappa>1$, let Δ be the critical subgraph $(\Pi \setminus J) \setminus A$ corresponding to Lemma 5.4, and let $\overline{\Lambda} = (\Lambda_\beta : \beta<\varkappa)$ be an impeding ladder for Π. Define $M_\alpha = M_\Delta \cap M_{\Lambda_\alpha}$, $K_\alpha = \cup\{M_\beta : \beta<\alpha\}$, $\Pi_\alpha = \vee\{\Lambda_\beta : \beta<\alpha\}$, and $A_\alpha = W_{\Pi_\alpha} \setminus W_\Delta$.

Let S_1, S_2, and S_3 respectively be the set of those $\alpha<\varkappa$ such that Λ_α has property (i), (ii), and (iii) from (I1). If $\delta \in S_1$ and $\delta<\alpha$, then $|A_\alpha \cap W_{\Lambda_\delta}| < \varkappa$ by Lemma 5.4. If $\delta \in S_3 \cap \alpha$, then $A_\alpha \cap W_{\Lambda_\delta} = \emptyset$, if

$\delta \in S_2 \cap \alpha$, then $|A_\alpha \cap W_{\Lambda_\delta}| = 1$. Consequently $|A_\alpha| = |\bigcup\{A_\alpha \cap W_{\Lambda_\delta} : \delta < \alpha\}| < \varkappa$, since $\varkappa$ is regular. If we can show that K_α is critical in F_{A_α}, then the sequence $(X_\alpha : \alpha < \varkappa)$, $X_\alpha = F(K_\alpha) \cup A_\alpha$, is a natural candidate for a $\varkappa$-test.

The construction of Δ shows that $\Delta \cap \Pi_\alpha$ is critical. Since Π_α is saturated in Γ_F, K_α is critical in $(F \restriction M_\Pi)_{A_\alpha}$ and consequently in F_{A_α}. Clearly $X_0 = \emptyset$ and $X_\alpha \subseteq X_\beta$ for $\alpha < \beta$. To show that the sequence $(X_\alpha : \alpha < \varkappa)$ is continuous, let $\alpha < \varkappa$ be a limit ordinal. Then

$$
\begin{aligned}
X_\alpha &= A_\alpha \cup F(K_\alpha) \\
&= (W_{\Pi_\alpha} \setminus W_\Delta) \cup F(\bigcup\{K_\beta : \beta < \alpha\}) \\
&= \bigcup\{W_{\Pi_\beta} \setminus W_\Delta : \beta < \alpha\} \cup \bigcup\{F(K_\beta) : \beta < \alpha\} \\
&= \bigcup\{(W_{\Pi_\beta} \setminus W_\Delta) \cup F(K_\beta) : \beta < \alpha\} \\
&= \bigcup\{A_\beta \cup F(K_\beta) : \beta < \alpha\} \\
&= \bigcup\{X_\beta : \beta < \alpha\}.
\end{aligned}
$$

Therefore $(X_\alpha : \alpha < \varkappa)$ is a $\varkappa$-test. We assume that neither condition (2) nor (3) nor (4) is satisfied and shall show that F fulfils condition (1). The assumption implies that there exists a club C in $\varkappa$ with $0 \in C$ (this results from the fact that (4) does not hold) such that $F_{X_\alpha}^{X_\beta}$ is ρ-unblocked for each $\alpha \in C$ and for any $\beta \geq \alpha$, $\rho < \varkappa$, and that further the following holds: If $\alpha \in C$ and $F(i) \subseteq X_\alpha$, then there exists a $\beta < \alpha$ with $F(i) \subseteq X_\beta$. The set $S^* = C \cap S_1$ is stationary. Suppose now for contradiction that there exists an $\alpha \in S^*$ such that, for each $i \in M_{\Lambda_\alpha}$, we have $F(i) \not\subseteq X_\alpha$. By construction, $W_{\Lambda_\alpha} \subseteq A_{\alpha+1} \cup F(K_{\alpha+1}) = X_{\alpha+1}$. If Σ is the bipartite graph corresponding to $F_{X_\alpha}^{X_{\alpha+1}}$, we can conclude that $\Lambda_\alpha \leq \Sigma$, and therefore Λ_α is a saturated subgraph of Σ, since Λ_α is saturated in $\Pi \setminus \Pi_\alpha$. So Λ_α is a ρ-impediment in Σ for some $\rho < \varkappa$. On the other hand Σ is ρ-unblocked and, by the hypothesis of the induction, it cannot contain a ρ-impediment. This contradiction shows that, for each $\alpha \in S^*$, there exist a $\beta_\alpha < \alpha$ and an $i_\alpha \in M_{\Lambda_\alpha}$ such that, by the choice of C, $F(i_\alpha) \subseteq X_{\beta_\alpha}$. Clearly $i_\alpha \neq i_\beta$ for $\alpha \neq \beta$. Now Fodor's Theorem yields a stationary set $S^{**} \subseteq S^*$ and a $\gamma < \varkappa$ such that $\beta_\alpha = \gamma$ for every $\alpha \in S^{**}$. This means $F(i_\alpha) \subseteq X_\gamma = F(K_\gamma) \cup A_\gamma$ for every $\alpha \in S^{**}$. So K_γ is critical in F_{A_γ} and $|\{i \in I \setminus K_\gamma : F(i) \subseteq F(K_\gamma) \cup A_\gamma\}| > |A_\gamma|$, which shows that condition (1) of Lemma 5.6 is satisfied. This completes the proof of the lemma.

The notion of an impediment allows an elegant proof of the following compactness theorem for families $F = (F(i) : i \in I)$ for which $|I|$ is singular.

Theorem 5.11. Let $F = (F(i) : i \in I)$ be a family, let $\lambda = |I|$ be singular, and let $\varkappa < \lambda$ be a cardinal such that $|F(i)| \leq \varkappa$ for each $i \in I$. If, for each subset J of I with $|J| < \lambda$, the family $F \upharpoonright J$ possesses a marriage, then there exists a marriage of F.

Proof. To get a contradiction, let us suppose that the family F satisfies the hypotheses of the theorem but possesses no marriage. Then, by Theorem 5.1, Γ_F contains a ρ-impediment Π. Let Δ be the critical subgraph of Π according to Lemma 5.4, and let h be a marriage of Δ. By property (I2), we have $|M_\Pi| \geq \rho$; consequently $\rho \leq \lambda$ and even $\rho < \lambda$, since ρ is regular. Define $J_0 = M_\Pi \setminus M_\Delta$, $A_0 = E[J_0]$, $J_{n+1} = h^{-1}[A_n]$, $A_{n+1} = E[J_n]$, and $J = \bigcup\{J_n : n \in \omega\}$. By Lemma 5.4, $|J_0| = \rho < \lambda$, and, since $|F(i)| \leq \varkappa < \lambda$ for every $i \in I$, we have $|A_0| < \mu := (\rho \cdot \varkappa)^+$. In the same way we get that $|J_n| < \mu$, $|A_n| < \mu$ for $n \in \omega$, hence $|J| < \lambda$. Let g be a marriage of $F \upharpoonright J$, which exists by assumption. There exists no $i \in M_\Pi \setminus J$ such that $h(i) \in \bigcup\{A_n : n \in \omega\}$, since otherwise i would be an element of J. Therefore $h \upharpoonright (I \setminus J) \cup g$ is a marriage of $F \upharpoonright M_\Pi$. Since Π is saturated in Γ_F, this yields immediately a matching of M_Π into W_Π, which contradicts Theorem 5.1.

The assumption "$|F(i)| \leq \varkappa$ for any $i \in I$ and some $\varkappa < \lambda$" cannot be dropped - let $F(0) = \aleph_\omega \setminus \{0\}$ and $F(\alpha) = \{\alpha\}$ for $0 < \alpha < \aleph_\omega$. Every proper subfamily of F has a marriage, but F possesses no marriage. But even the assumption "$|F(i)| < \lambda$ for any $i \in I$" is not sufficient to prove the theorem: Let $I = \aleph_\omega \cup \{-1\}$, $G(-1) = \{\aleph_n : n \in \omega\}$, $G(\aleph_n) = \{\beta : \aleph_n \leq \beta < \aleph_{n+1}\}$ for $n \in \omega$, and $G(\alpha) = \{\alpha\}$ otherwise. It is easy to see that, for every $i \in I$, the family $G \upharpoonright I \setminus \{i\}$ has a marriage, and that G possesses no marriage.

The following lemma yields a stronger version of Theorem 5.11: The theorem is valid for families $F = (F(i) : i \in I)$ such that $F \upharpoonright (I \setminus L)$ has a marriage for some $L \subseteq I$ of cardinality λ. Of course $L = I$ then yields Theorem 5.11. Lemma 5.12 yields just the proof of the necessity of the criterion in [ANS1], i.e. we reprove one direction of 5.1.

Lemma 5.12. If Π is a ρ-impediment and if L is a subset of M_Π such that $\Pi\setminus L$ is matchable, then $\rho\leq|L|$.

Proof. We use transfinite induction on ρ. If $\rho = 1$ and $\Pi\setminus\{m\}$ is critical for some $m\in M_\Pi$, then L cannot be empty, hence $\rho\leq|L|$. Now let $\rho>1$ and assume for contradiction that $\Pi\setminus L$ is matchable for some $L\subseteq M_\Pi$ with $|L|<\rho$. Let h be a matching of $\Pi\setminus L$, Λ be an obstructive ladder for Π, $S_1 = \{\alpha<\rho : \Lambda_\alpha$ is a μ-impediment for some $\mu<\rho\}$, and $D = \{\alpha<\rho : L\cap M_{\Lambda_\alpha} \neq \emptyset\}$. Since $|L|<\rho$ and S_1 is stationary in ρ by property (I2), the set $S = S_1\setminus D$ is stationary, too. Especially $M_{\Lambda_\alpha}\subseteq M_\Pi\setminus L$ for each $\alpha\in S$. Since the induction hypothesis implies that Λ_α is unmatchable, there must exist, for each $\alpha\in S$, an $m_\alpha\in M_{\Lambda_\alpha}$ and a $\varphi(\alpha)<\alpha$ such that $h(m_\alpha)\in W_{\Lambda_{\varphi(\alpha)}}$. Apply Fodor's Theorem to get an ordinal β with $|\varphi^{-1}[\{\beta\}]| = \rho$. Take $\Sigma = \Pi_{\beta+1} = \vee\{\Lambda_\gamma : \gamma<\beta+1\}$ and $B = \{h(m_\alpha) : \varphi(\alpha) = \beta\}$. Then $h\restriction M_\Sigma$ is a matching of $\Sigma\setminus(L\cap M_\Sigma)\setminus B$. Further the proof of Lemma 5.4 shows that $\Sigma\setminus J^*\setminus A^*$ is critical for some sets $J^*\subseteq M_\Sigma$, $A^*\subseteq W_\Sigma$ with $|A^*\cup J^*|<\rho$. Theorem 2.10, applied to the family F_Σ, now yields

$$|B| + |(M_\Sigma\setminus L)\setminus(M_\Sigma\setminus J^*)| \leq |A^*| + |(M_\Sigma\setminus J^*)\setminus(M_\Sigma\setminus L)|.$$

We have $|B| = \rho$ and $|A^*|<\rho$. Further $|(M_\Sigma\setminus J^*)\setminus(M_\Sigma\setminus L)|\leq|L|<\rho$, hence $|B|<\rho$. This contradiction completes the proof of Lemma 5.12.

Corollary 5.13. Let $F = (F(i) : i\in I)$ be a family such that $F\restriction I\setminus L$ has a marriage for some subset L of I of cardinality λ and such that F satisfies the hypotheses of Theorem 5.11. Then F possesses a marriage.

Proof. We go back to the proof of Theorem 5.11. Since the ρ-impediment Π has the property that $\Pi\setminus L$ has a marriage, Lemma 5.12 implies $\rho\leq|M_\Pi\cap L|\leq\lambda$. Again we conclude $\rho<\lambda$, since λ is singular, and the rest of the proof remains the same.

§6. The Duality Theorem and its applications

The main result in Aharoni's paper [A3] concerning the aim of this book is the following theorem:

<u>Theorem 6.1</u>. If $\Gamma = (M,W,E)$ is a bipartite graph, then Γ has no marriage if and only if there exists a subset X of W such that X is matchable into $D_\Gamma(X)$ and $D_\Gamma(X)$ is unmatchable.

Please remember that $D_\Gamma(X) = \{m \in M : E<m> \subseteq X\}$ is the demand of X, i.e. the set of those men who only know women in X. We are going to prove this theorem in the following section. In this section some important applications are presented. The first is the proof of a strong version of König's Duality Theorem (Theorem 6.2), which initiated the proof of Theorem 6.1 in [A3]. In fact, the two properties of bipartite graphs in Theorem 6.1 and Theorem 6.2 are seen to be equivalent in such a natural way that we can interpret the Duality Theorem as a marriage criterion or Theorem 6.1 as a reformulation of the Duality Theorem. With this theorem we can give Aharoni's proof of Menger's Theorem for graphs without infinite paths and, additionally, we can generalize the proof of Oellrich and Steffens in [OS] for Dilworth's decomposition theorem for arbitrary partially ordered sets without an infinite chain. Readers who want to read first the proof of Theorem 6.1 may omit this section without loss of any information which would be necessary for the study of the following sections.

<u>Definition</u>. Let $\Gamma = (M,W,E)$ be a bipartite graph. A set $\mathcal{O} \subseteq M \cup W$ is called a <u>cover of Γ</u> if $E \subseteq \mathcal{O} \otimes (M \cup W)$, i.e. if every edge contains a vertex of $\mathcal{O}$.

We can now state the most important theorem of this section, which was called König's Duality Theorem by Aharoni:

<u>Theorem 6.2</u> If $\Gamma = (M,W,E)$ is a bipartite graph, then there exist a matching f in Γ and a cover $\mathcal{O}$ such that $\mathcal{O}$ consists of a choice of precisely one vertex from each edge in f.

<u>Proof</u>. If Γ has a marriage g, then our theorem is true for $f = g$ and $\mathcal{O} = M$. So let Γ have no marriage. Choose $X_0 \subseteq W$ with the properties in Theorem 6.1, i.e. X_0 is matchable by f_0 into $D_\Gamma(X_0)$, but $D_\Gamma(X_0)$ has no matching. Take $\Gamma_1 = (\Gamma \setminus D_\Gamma(X_0)) \setminus X_0$. If we can find a cover $\mathcal{O}^*$ and a matching f^* for Γ_1 with the required properties, then $\mathcal{O} = \mathcal{O}^* \cup X_0$ and $f = f^* \cup f_0$ are fulfilling our claims, as it is easy to verify. So it

suffices to investigate Γ_1. Since we can proceed with Γ_1 in the same way as with $\Gamma_0 := \Gamma$, the following inductive definition is motivated:

Let Γ_β, X_β and f_β be defined for each $\beta < \alpha$, where $X_\beta \subseteq W_{\Gamma_\beta} =: W_\beta$ and f_β is a matching of X_β into $D_{\Gamma_\beta}(X_\beta) =: D_\beta(X_\beta)$.

If $\alpha = \sigma + 1$ is a successor ordinal, then let $\Gamma_\alpha = (\Gamma_\sigma \setminus D_\sigma(X_\sigma)) \setminus X_\sigma$. Choose X_α and f_α according to Theorem 6.1, applied to Γ_α, if Γ_α has no marriage. Otherwise take f_α as a marriage of Γ_α and $X_\alpha = M_{\Gamma_\alpha} =: M_\alpha$, and terminate the definition.

If α is a limit ordinal, then let

$$\Gamma_\alpha = (\Gamma \setminus \cup\{D_\beta(X_\beta) : \beta < \alpha\}) \setminus \cup\{X_\beta : \beta < \alpha\}$$

and proceed in the same way as in the successor step.

If our definition does not stop at α, then we have $X_\alpha \neq \emptyset$ or $D_\alpha(X_\alpha) \neq \emptyset$. Since $M \cup W$ is a set, the sequence of bipartite graphs has a last element, say Γ_σ. Now define $f = \cup\{f_\beta : \beta \leq \sigma\}$ and $\mathcal{O} = M_\sigma \cup \cup\{X_\beta : \beta < \sigma\}$. It is obvious that f is a matching, $\mathcal{O} \subseteq f[M \cup W]$, and that every edge of f contains exactly one element of $\mathcal{O}$. Further it is not difficult to show that every edge of Γ contains an element of $\mathcal{O}$. So our theorem is proved as a consequence of Theorem 6.1.

It is easy to see that Theorem 6.2 implies Theorem 6.1: If Γ has no marriage and f and $\mathcal{O}$ are the matching and the cover yielded by Theorem 6.2, take $X = \mathcal{O} \cap W$. Since $\mathcal{O}$ is a cover, there is no edge from $M \setminus \mathcal{O}$ to $W \setminus \mathcal{O}$ and so $M \setminus \mathcal{O} \subseteq D_\Gamma(X)$. Further there exists no edge of f from $\mathcal{O} \cap M$ to X. Consequently $D_\Gamma(X) = M \setminus \mathcal{O}$ is unmatchable, since such a matching h would yield the marriage $g = h \cup f \restriction (\mathcal{O} \cap M)$ of Γ. Further f is a matching of X into $D_\Gamma(X)$, since $f[X] \subseteq M \setminus \mathcal{O}$. So the nontrivial direction of Theorem 6.1 is proved.

An immediate consequence of Theorem 6.2 is the following theorem, which is known in a weaker version as Dilworth's Theorem. First we need a

Definition. If $(Q,<)$ is a partial ordering, then we call a subset $A \subseteq Q$ an antichain if the elements of A are pairwise incomparable with respect to <. Remember that a chain is a linearly ordered subset of Q.

Theorem 6.3. If $(Q,<)$ is a partial ordering without infinite chains, then there exist a partition $(C_i : i \in I)$ of Q into pairwise disjoint chains and an antichain $A \subseteq Q$, such that A contains one element of each chain C_i, $i \in I$.

Proof. Let $h : Q \to Q'$ be a bijection, where $Q \cap Q' = \emptyset$, and let $\Gamma = (Q,Q',E)$ be the bipartite graph with edge set $\{\{p,h(q)\} : p,q \in Q$ and $p<q\}$. We write p' for $h(p)$. Let f and $\mathcal{O}$ be the matching and the cover for Γ according to Theorem 6.2. Define $\overline{f}(p) = p$ if there is no $q \in Q$ such that the edge $\{p,q'\}$ belongs to f, and $\overline{f}(p) = q$ if $\{p,q'\} \in f$, and call $q \in Q$ an initial point if there is no $p \neq q$ with $\overline{f}(p) = q$. Take J as the set of all initial points. For every $p \in J$ we define a sequence $(p_n : n \in \omega)$ as follows: $p_0 = p$ and $p_{n+1} = \overline{f}(p_n)$. Since $(Q,<)$ has no infinite chain, the set $C_p = \{p_n : n \in \omega\}$ is finite. It is easy to see that $(C_p : p \in J)$ is a partition of Q into pairwise disjoint, nonempty chains.

Now we define the antichain A by $A = \{\overline{p} : p \in J\}$, where $\overline{p} = p_{n(p)}$ and $n(p)$ is the least natural number n such that $p_n \notin \mathcal{O}$. To show that A is an antichain, assume for contradiction that we have $\overline{p} < \overline{q}$ for some $p,q \in J$. Then $\{\overline{p},\overline{q}'\} \in E$, and this edge contains an element of the cover $\mathcal{O}$. By definition, this cannot be $\overline{p}$, hence $\overline{q}' \in \mathcal{O}$ and consequently $\{r,\overline{q}'\} \in f$ for some $r \in Q$ by the properties of f and $\mathcal{O}$. We have $r \neq \overline{q}$ and $\overline{f}(r) = \overline{q}$, so $\overline{q}$ is not an initial point. So, by the definition of $\overline{q}$, we have $r \in \mathcal{O}$. But now the edge $\{r,\overline{q}'\} \in f$ contains two elements of $\mathcal{O}$, which contradicts the properties of f and $\mathcal{O}$.

The rest of this section is devoted to the proof of Menger's Theorem for graphs without infinite paths.

Let $G = (V,E)$ be an undirected graph or a directed graph, i.e. $E \subseteq V \times V$, and let A and B be subsets of V. The notion of a path in a directed graph can be defined in the same way as in the undirected case. An A-B-path is a finite path $(v_i : i \leq n)$ such that $v_0 \in A$ and $v_n \in B$. If p is a path in G (possibly one-way infinite or two-way infinite), then we say that p is A-B-directed if (1) either it has no first vertex or its first vertex is an element of A and (2) either it has no last vertex or its last vertex is an element of B. A subset S of V is called A-B-separating if every A-B-path contains at least one vertex from S.

Menger proved that every finite graph (V,E) has the following property (where "disjoint" means "vertex-disjoint").

(Me) For any pair (A,B) of subsets of V, there exist a family $\mathcal{F}$ of disjoint A-B-paths and an A-B-separating set S such that S consists of a choice of precisely one vertex from each path in $\mathcal{F}$.

P. Erdös made the conjecture that property (Me) holds for every infinite graph. In [PS3] Podewski and Steffens proved this conjecture for countable graphs without infinite paths.

To be precise, they proved the following stronger version of the property (Me*) in Theorem 6.5 for countable graphs: For any pair (A,B) of subsets of V, there exists a family $\mathcal{F}$ of pairwise disjoint paths which either are A-B-paths or are infinite and start in A, and there exists an A-B-separating set S such that S consists of a choice of precisely one vertex from each path in $\mathcal{F}$. [1)]

We now give Aharoni's proof ([A2]) of the following theorem.

Theorem 6.4. Property (Me) holds for any graph that does not contain an infinite path.

Aharoni proved a stronger result which obviously implies Theorem 6.4:

Theorem 6.5.
Every (directed or undirected) graph G = (V,E) has the following property:

(Me*) For any pair (A,B) of subsets of V, there exist a family $\mathcal{F}$ of pairwise disjoint, A-B-directed paths and an A-B-separating set S such that S consists of a choice of precisely one vertex from each path in $\mathcal{F}$.

Proof. We give the proof for directed graphs. If G is undirected, then we get the result for G from this proof as follows. Each edge $\{u,v\} \in E$ is replaced by the four directed edges (u,x), (x,v), (v,y), and (y,u), where $x = x(u,v)$ and $y = y(u,v)$ are new vertices, such that for different edges of G we add different vertices. Assume that the resulting directed graph $G' = (V',E')$ has property (Me*) via $\mathcal{F}'$ and S'. Let $p \in \mathcal{F}'$. Replace each subpath (v,y,u)

1) In July 1986, the authors received a preprint of R.Aharoni in which he proves (Me) for arbitrary countable graphs.

by (v,u) and each subpath (u,x,v) by (u,v), and let p^* be the resulting path. p^* is an A-B-directed path in G, and $\{p^*: p\in\mathcal{F}'\}$ is a family of pairwise disjoint, A-B-directed paths. If $x(u,v)\in S'$, then there exists exactly one $p\in\mathcal{F}'$ such that $x(u,v)\in p$. Hence p must contain u and v, since it is A-B-directed. Replace $x(u,v)$ in S' by u; analogously proceed with $y(u,v)$, leave the vertices of $V(G)\cap S'$, unchanged, and denote the set which results from S' by S. If q is an A-B-path in G, then it is clear how to obtain a corresponding A-B-path p in G'. It contains an element of S', and, by our construction of S, $p^* = q$ contains an element of S.

We further assume that the following two conditions are satisfied for our directed graph G:

(1) $A\cap B = \emptyset$.

(2) There exists no edge $(u,v)\in E$ such that $u\in B$ or $v\in A$.

We show that this is no loss of generality. If our result is true for graphs which satisfy condition (1) and G is arbitrary, let $G' = G\setminus A\cap B$ and let $\mathcal{F}'$ and S' guarantee (Me^*) for G'. Then $\mathcal{F} = \mathcal{F}'\cup\{(x): x\in A\cap B\}$ and $S = S'\cup(A\cap B)$ fulfil our claims for G. If our result is true for graphs which satisfy conditions (1) and (2) and if G satisfies (1), then let G' be the graph with edge set $E = E(G)\setminus\{(u,v): u\in B \text{ or } v\in A\}$, and let $\mathcal{F}$ and S guarantee (Me^*) for G'. It remains to show that every A-B-path $p = (v_k: k<n)$ in G contains an element of S. If p has an edge (u,v) with $v\in A$, then there is a last edge (v_l,v_{l+1}) of this kind, since p ends in B and $A\cap B = \emptyset$. Analogously there is a first edge (v_j,v_{j+1}) of p with $j\geq l$ and $v_j\in B$, if there is some edge of this kind. If (v_l,v_{l+1}) and (v_j,v_{j+1}) exist, then the path $(v_k: l+1\leq k\leq j)$ is an A-B-path in G' and contains a vertex of S (the other cases are treated in the same way). So S is A-B-separating in G.

Consequently we can assume for the rest of the proof that $G = (V,E)$ is a directed graph which satisfies properties (1) and (2).

Since we want to apply Theorem 6.2, we assign to G a bipartite graph Γ in the following way. Let V' and V" two disjoint sets and $h_1: V\to V'$ and $h_2: V\to V''$ two bijections. We write v' for $h_1(v)$ and v" for $h_2(v)$. Analogously let $W' = h_1[W]$ and $W'' = h_2[W]$ for every $W\subseteq V$. Define $X = V\setminus(A\cup B)$ and $h((u,v)) = \{u',v''\}$ for every $(u,v)\in E$. Now take

$$\Gamma = (A'\cup X', B''\cup X'', h[E]\cup\{\{x',x''\}: x\in X\}).$$

By Theorem 6.2, there exist a matching f_1 and a cover $\mathcal{O}$ such that $\mathcal{O}$

consists of a choice of precisely one vertex from each edge in f_1. Take $f = f_1 \cap h[E]$ and $g = h^{-1} \mid f$.

We give a description of the proof. The first idea is to choose the set of all A-B-directed paths whose edges are corresponding to edges of the matching f_1, i.e whose edges are in $g[f]$. Obviously these paths are pairwise disjoint. Let $p = (v_i : k_1 < i < k_2)$ be such a path $(-\infty \le k_1 \le 0 \le k_2 \le \infty)$.

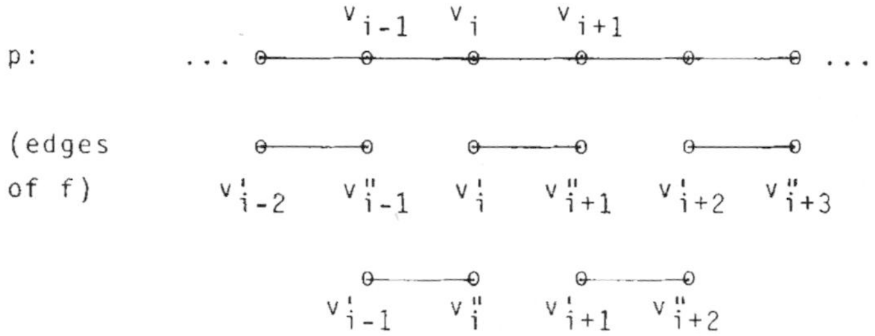

p: ... v_{i-1} v_i v_{i+1} ...

(edges of f) v'_{i-2} v''_{i-1} v'_i v''_{i+1} v'_{i+2} v''_{i+3}

v'_{i-1} v''_i v'_{i+1} v''_{i+2}

We always have $v'_i \in \mathcal{O}$ or $v''_{i+1} \in \mathcal{O}$. Further the edges (x', x'') in Γ guarantee the property

(3) $x' \in \mathcal{O}$ or $x'' \in \mathcal{O}$ for every $x \in X$,

since $\mathcal{O}$ is a cover.

Now consider a vertex $v_i \in X$ such that $v'_i \in \mathcal{O}$ <u>and</u> $v''_i \in \mathcal{O}$. Since $\{v'_i, v''_{i+1}\} \in f$, we have $v''_{i+1} \notin \mathcal{O}$, hence $v'_{i+1} \in \mathcal{O}$, $v'_{i+2} \in \mathcal{O}$ and so on. Further there is no $v = v_{i+k}$, $k \ge 1$, such that $v' \in \mathcal{O}$ and $v'' \in \mathcal{O}$. Analogously for no $u = v_{i-k}$, $k \ge 1$, we have $u' \in \mathcal{O}$ and $u'' \in \mathcal{O}$. Therefore the path p contains at most one vertex $v_i \in X$ with $v'_i \in \mathcal{O}$ and $v''_i \in \mathcal{O}$. We call any vertex $v_i \in X$ with the property that $v'_i \in \mathcal{O}$ and $v''_i \in \mathcal{O}$ <u>doubly marked</u>. Likewise a vertex u with $u \in A$ and $u' \in \mathcal{O}$ or with $u \in B$ and $u'' \in \mathcal{O}$ is called <u>doubly marked</u>. If p contains a vertex with $u \in A$ and $u' \in \mathcal{O}$, then, by property (2), $u = v_{k_1+1}$ is the first vertex of p and, as above, for every $i > k_1 + 1$ we have $v''_i \notin \mathcal{O}$. Analogously, if p contains $v \in B$ with $v'' \in \mathcal{O}$, then $v'' = v_{k_2-1}$ is the last vertex of p and for every $i < k_2 - 1$ we have $v_i' \notin \mathcal{O}$. Thus we have proved the following

<u>Claim 1</u>.
If p is a path whose edges are in $g[f]$, then p contains at most one doubly marked vertex.

Next we observe that each A-B-path $p = (v_i : 0 \le i \le n)$ contains a doubly marked vertex: let $v_0 = a$, $v_n = b$. If $a' \notin \mathcal{O}$, then $v''_1 \in \mathcal{O}$. If v_1 is not doubly marked, then $v''_2 \in \mathcal{O}$ and so on. Finally we get that if neither a nor any v_i is doubly marked for $i < n$, then $v''_n \in \mathcal{O}$, hence

$v_n = b$ is doubly marked. If we choose $\mathcal{P}$ as the set of all A-B-directed paths whose edges are in $g[f]$ and which contain a doubly marked vertex, then we only have to show that, for every doubly marked vertex, there exists an A-B-directed path in $\mathcal{P}$ which contains it, and this will complete the proof of Theorem 6.5 with S as the set of doubly marked vertices.

Claim 2.

If $v \in V$ is doubly marked, then there exists an A-B-directed path which contains v and has all its edges in $g[f]$.

We construct vertices v_k and v_{-k} for every $k \geq 0$, such that (v_z, v_{z+1}) belongs to $g[f]$ for each integer z. Clearly $v_0 = v$. If $v_0 \in B$, then the construction of the sequence $(v_k : k \geq 0)$ is finished. Let $v_0 \notin B$. Then $v_0' \in \mathcal{O}$ and further $v_0'' \in \mathcal{O}$ or $v_0 \in A$, since v_0 is doubly marked. Therefore there exists an edge $(v_0', v_1'') \in f_1$ for some $v_1 \in V$. We have $(v_0', v_1'') \in f$: If $v_0 \in A$, this is clear. Let $v_0 \notin A$. Then $v_0'' \in \mathcal{O}$, hence (v_0', v_1'') cannot be the edge (v_0', v_0'') which would contain two elements of $\mathcal{O}$, and consequently $(v_0', v_1'') \in f$. If $v_1 \in B$, we stop the construction. Otherwise $v_1 \in X$ and, by (3), $v_1' \in \mathcal{O}$ or $v_1'' \in \mathcal{O}$. Since $v_0' \in \mathcal{O}$ and $(v_0', v_1'') \in f$, we have $v_1' \in \mathcal{O}$. Now assume that we have defined $v_0, v_1, \ldots, v_k$ $(k \geq 1)$, such that for each $i < k$ we have $(v_i', v_{i+1}'') \in f$ and $v_{i+1}' \in \mathcal{O}$. Since $v_k' \in \mathcal{O}$, we can choose v_{k+1} such that $(v_k', v_{k+1}'') \in f_1$. Again we show that $v_k \neq v_{k+1}$, i.e. $(v_k', v_{k+1}'') \in f$. If $v_k = v_{k+1}$, we would have $(v_k', v_k'') \in f_1$ and $(v_{k-1}', v_k'') \in f \subseteq f_1$, and this would be a contradiction, since f_1 is a matching. Consequently $(v_k', v_{k+1}'') \in f$ and $v_{k+1}'' \notin \mathcal{O}$. If $v_{k+1} \in B$, we finish the definition. If $v_{k+1} \notin B$, then $v_{k+1} \in X$ by (2), and by (3) we get $v_{k+1}' \in \mathcal{O}$. Note that $v_{k+1} \neq v_0$, since $v_{k+1}'' \notin \mathcal{O}$, and $v_{k+1} \neq v_i$ for $0 < i < k$, since f is a matching. The definition of the injective sequence $(v_k : k \geq 0)$ is completed, and this path is infinite or ends in B, and every edge of it belongs to $g[f]$. In the same way we define the injective sequence $(v_k : k \leq 0)$ such that $(v_i, v_{i+1}) \in g[f]$ and $v_i'' \in \mathcal{O}$ or $v_i \in A$ for each $i < 0$. So the path $(v_k : k \leq 0)$ is infinite or starts in A, and every edge of it belongs to $g[f]$. If $v_k = v_{-i}$ for some $i, k > 0$, choose the least $i > 0$ with the property that there exists $k > 0$ with $v_k = v_{-i}$. Then $v_{-i-1} \neq v_{k-1}$ and $(v_{-i-1}', v_k''), (v_{k-1}', v_k'') \in f$, a contradiction. Consequently the concatenation of our two sequences yields an A-B-directed path which contains v and has its edges in $g[f]$.

§7. Proof of Aharoni's criterion

The proof of the criterion which led to the proof of the duality theorem is rather extensive. Therefore we give the definition step by step and try to develop some ideas which yield the red thread for the understanding of the proof. We formulate the criterion once more:

Theorem 7.0. If $\Gamma = (M,W,E)$ is a bipartite graph, then Γ possesses no marriage if and only if there exists a subset X of W such that X is matchable into the demand $D_\Gamma(X)$ of X in Γ and $D_\Gamma(X)$ is unmatchable in Γ.

One direction of the theorem is obvious: If $D_\Gamma(X)$ is unmatchable for some $X \subseteq W$, then Γ is unmatchable. Therefore let us assume that Γ has no marriage. Then, by Theorem 5.1 and Lemma 5.3, Γ contains a $\varkappa$-obstruction Π for some $\varkappa \in Rc$. If we can find a set X with the desired properties as a subset of W_Π, then we have $D_\Gamma(X) \cap M_\Pi = D_\Pi(X)$, since Π is saturated in Γ. Hence it suffices to prove the theorem for obstructions, for if $D_\Pi(X)$ is not matchable, then all the more this holds for $D_\Gamma(X)$. Consequently let $\Gamma = (M,W,E)$ for the rest of the whole section be a $\varkappa$-obstruction. If $\varkappa = 1$, then $\Gamma \setminus \{m\}$ is critical for some $m \in M$ and it is easy to see that we can choose $X = W_\Gamma$. Therefore let $\varkappa > 1$ and assume that the assertion is true for every ρ-obstruction with $\rho < \varkappa$. Let $\bar{\Lambda} = (\Lambda_\alpha : \alpha < \varkappa)$ be an obstructive $\varkappa$-ladder for Π, let $\Lambda_\alpha = (M_\alpha, W_\alpha, E_\alpha)$, and let $S = \{\alpha < \varkappa : \Lambda_\alpha$ is a ρ-obstruction for some $\rho < \varkappa\}$. S is stationary by property (O2) of an obstruction.

The hypothesis of the induction yields, for each $\alpha \in S$, a subset X_α of W_α such that $D_{\Lambda_\alpha}(X_\alpha)$ is unmatchable in Λ_α and X_α is matchable into $D_{\Lambda_\alpha}(X_\alpha)$. Let h_α be the matching from X_α into $D_{\Lambda_\alpha}(X_\alpha)$. We shall see later on that the inductive proof requires a stronger version of Theorem 7.0. If we take first a naive attitude, then certainly $h^* = \bigcup\{h_\alpha : \alpha \in S\}$ is a matching from $X^* = \bigcup\{X_\alpha : \alpha \in S\}$ into M.

But in general the following may occur:

(1) h^* is not a matching into $D_\Gamma(X^*)$.

Namely if $\alpha \in S$ and if the man $m \in h_\alpha[W_\alpha]$ knows a woman w such that $\Lambda_\beta = (\emptyset, \{w\}, \emptyset)$ for some $\beta < \alpha$, then $w \notin X^*$, and consequently we have $m \in h^*[X^*]$, but $m \notin D_\Gamma(X^*)$.

(2) The demand of X^* in Γ is matchable.

For example choose $\Lambda_\alpha = (\{i_\alpha\},\emptyset,\emptyset)$ for limit ordinals $\alpha<\kappa$ and take care that every man i_α knows some woman - this is always possible with help of the rungs $\Lambda_{\beta+1} = (\emptyset,\{w_{\beta+1}\},\emptyset)$, $\beta<\alpha$. Then $D_{\Lambda_\alpha}(\emptyset) = \{i_\alpha\}$ is unmatchable in Λ_α but $D_\Gamma(\emptyset) = \emptyset$ is matchable in Γ.

Nevertheless we will use the matchings h_α to define a matching j in Γ, which enables us to define a subset X of W which satisfies our claims (- j will not be the marriage of X). An essential property will be that, for each $\alpha\in S$, there are less than κ women in W_α which do not belong to an edge of j. This is obvious for $\kappa = 1$. If $\kappa>1$, we succeed by defining, for every ρ-obstruction Π, a critical subgraph $\Delta(\Pi)$ and a marriage $f(\Pi)$ of $\Delta(\Pi)$ such that $|(W_\Pi \cup M_\Pi)\setminus(W_\Delta \cup M_\Delta)| \leq \rho$. In the induction proof we can use the additional property that the matching h_α from X_α into $D_{\Lambda_\alpha}(X_\alpha) =: L_\alpha$ satisfies $|h_\alpha \setminus f(\Lambda_\alpha)| \leq \rho_\alpha$ for some $\rho_\alpha<\kappa$, if Λ_α is a ρ_α-obstruction, and of course we have to show $|h\setminus f(\Gamma)| \leq \kappa$ for our matching h of X - this is the stronger version of Theorem 7.0. The property of $D_\Gamma(X)$ of being unmatchable will be proved as follows. The set of $\alpha\in S$ such that $L_\alpha = D_{\Lambda_\alpha}(X_\alpha)$ is a subset of $D(X)$ will be stationary in κ(Lemma 7.11). Since, for every $\alpha\in S$, L_α is not matchable in Λ_α, the supposition that $D(X)$ has a matching g in Γ yields, for each $\alpha\in S$, a man $m_\alpha \in L_\alpha$ who is matched by g to a woman of $\Lambda_{\beta(\alpha)}$ for some $\beta(\alpha)<\alpha$ (please remember: the subgraphs $\Pi_\alpha = \vee\{\Lambda_\beta : \beta<\alpha\}$ are saturated in Γ, therefore the men in Λ_α only know women in Λ_β for $\beta\leq\alpha$). By Fodor's Theorem, there exists some $\beta<\kappa$ such that $g(m_\alpha)\in W_\beta$ for all $\alpha\in S'$, where S' is a stationary subset of S. This will lead to a contradiction by arguments which result from the construction of X. We therefore concentrate our motivation on this construction and on the construction of a matching h of X into $D(X)$. Let $f_\alpha = f(\Lambda_\alpha)$ be the marriage of the critical subgraph $\Delta(\Lambda_\alpha)$ of Λ_α (we will have $f(\Gamma)\restriction\Lambda_\alpha = f_\alpha$). We take as j_α the union of h_α and the set of those edges in f_α which are compatible with h_α, i.e. $j_\alpha = h_\alpha \cup (f_\alpha \cap (M_\alpha \setminus h_\alpha[X_\alpha]) \otimes (W_\alpha \setminus X_\alpha))$. Finally we take $j = \cup\{j_\alpha : \alpha\in S\}$.

The most important concept is the notion of a j-alternating path. Such a path $g = (v_i : i\leq n)$ is always finite, it starts with an edge in $E\setminus j$ at some $v_0 \in W$, has edges alternately in j and $E\setminus j$ and ends with an edge in $E\setminus j$ at some $v_n \in M$.

If this man v_n is not contained in any edge of j, it is easy to see how to find a matching which matches additionally the woman v_0 from which p starts.

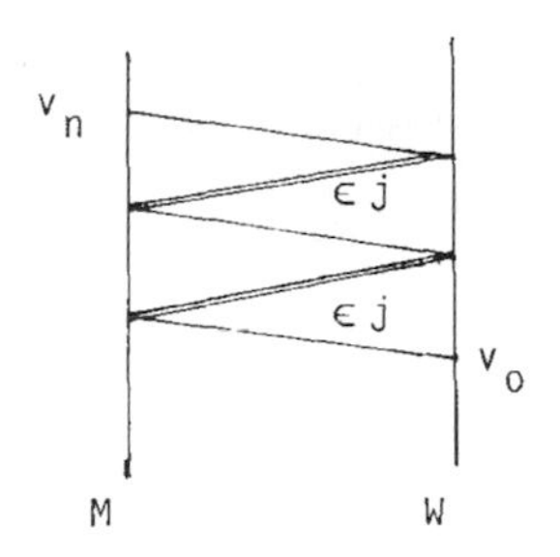

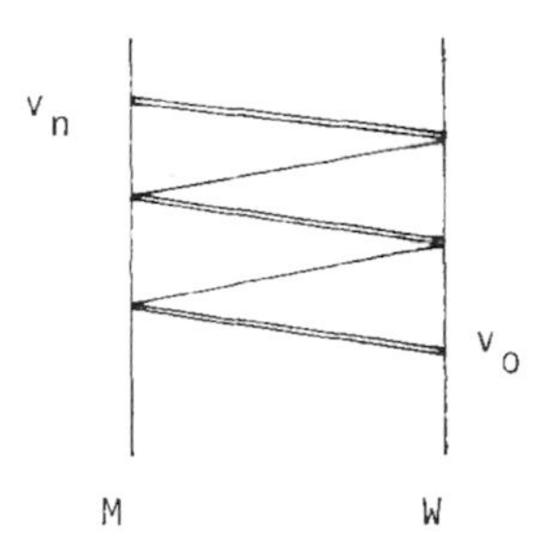

Instead of a j-alternating path we will speak of a j-path for the rest of this section. We are interested in those women for which there is a good chance to marry them additionally with help of j-paths. This motivates the following definition.

Definition. A family $(T_i : i \in I)$ of paths in a graph $G = (V,E)$ is called x-disjoint if $x \in V$ and if, for any $i,j \in I$ with $i \neq j$, the paths T_i and T_j have exactly the vertex x in common. It is called disjoint if its paths are pairwise disjoint. If Γ is our bipartite graph, then a family $(T_i : i \in I)$ of paths in Γ is called stationary if the endpoints of the paths are elements of different $L_\alpha := D_{\Lambda_\alpha}(X_\alpha)$ and if the set $\{\alpha \in S : \text{there exists an } i \in I \text{ such that } T_i \text{ ends in } L_\alpha\}$ is stationary in $\varkappa$. Finally we call an $x \in W$ popular if there exists a stationary, x-disjoint family $(T_i : i \in I)$ of j-paths such that every T_i starts in x.

Lemma 7.1. If $x \in W_\alpha$ is not an element of $j[M]$, and if the j-path T leads from x to L_β, i.e. its last vertex lies in L_β, then we have $\alpha < \beta$.

Proof. As it is known there are no edges from any L_γ to some W_σ with $\sigma > \gamma$. So we have $\alpha \leq \beta$, since the edges of j lead from any rung to the same rung of the ladder $\overline{\Lambda}$. Assume $\alpha = \beta$ and let $\{w_n, m_n\}$ be the last edge of $T = (w_0, m_0, \ldots, w_n, m_n)$. Then $m_n \in L_\beta = L_\alpha$, therefore $w_n \in E_{\Lambda_\alpha}\langle m_n \rangle \subseteq X_\alpha \subseteq j[M]$. If $n = 0$, this is a contradiction. Otherwise $\{m_{n-1}, w_n\} \in j$, so $m_{n-1} \in L_\alpha$, since $j[X_\alpha] \subseteq L_\alpha$. In the same way we get $w_{n-1} \in j[M]$ and finally $x \in j[M]$, a contradiction.

Lemma 7.1 is easy to prove, but it is important and should give the intuition that from any popular woman there are starting "stationary many" j-paths which end in different <u>higher</u> rungs of the ladder.

The principle for matching the elements of X in $\overleftarrow{\Gamma}(X,D(X))$ is the following: Each $x \in X$ is to be married with the help of two properties: x is popular, or $x \in j[M]$. If x is popular and if there is no j-edge from x into D(X), then x will be married by the method of j-paths. Not considering the requirement that the end-vertices of these paths possibly must be set free, the remaining vertices in $j[M] \cap X$ can keep their j-edges. When we are going to define the set X, this principle forces us to remove first the women who are neither popular nor elements of j[M] - call the set of these women R_0. The men in $E[R_0]$ now must be removed, too since they do not belong to the demand of the remaining women. Let $U_1 = E[R_0]$. Obviously we have $D_\Gamma(W \setminus R_0) = M \setminus U_1$. Now we have got new women who cannot be married by our principle, namely those women in $N_1 = j[U_1]$ which are not popular, since we have deleted their j-edges. If R_1 is the set of these women and $U_2 = E[R_1]$, we again must remove U_2 and get $D_\Gamma((W \setminus R_0) \setminus R_1) = (M \setminus U_1) \setminus U_2$. We see that, having fixed our principle, the rest of the procedure is inevitable, and we come to $X = W \setminus \cup\{R_i : i \in \omega\}$ and $J = D_\Gamma(X) = M \setminus \cup\{U_i : i \in \omega\}$.

<u>Definition</u>. Let P be the set of those $x \in W$ which are popular, let $Q = W \setminus P$ and $N_0 = W \setminus j[M]$. Define $R_0 = N_0 \cap Q$, $Z_0 = N_0 \cap P$, $U_0 = \emptyset$, and $U_k = E[R_{k-1}]$, $N_k = j[U_k]$, $R_k = N_k \cap Q$, $Z_k = N_k \cap P$ for $k \in \omega$, $k \neq 0$. Now take $R = \cup\{R_k : k \in \omega\}$, $Z = \cup\{Z_k : k \in \omega\}$, $U = \cup\{U_k : k \in \omega\}$, and define $X = W \setminus R$ and $J = M \setminus U$.

<u>Lemma 7.2</u>. $D_\Gamma(X) = J$.

The difficulties in the proof arise from the problem of matching the elements of Z, those are first the popular women leaving unmatched by j as elements of Z_0 and second those popular women whose j-edges we have removed by our construction. The method of j-paths presents, for every $x \in Z$, a choice of $\varkappa$ men, namely the endpoints of those paths which make x popular. The prerequisite is that they are at our disposal, i.e. that they are not blocked by other j-edges. We prevent this by a technical trick (Lemma 7.7) and remove from Λ_α, for suitable α's in S, a maximal critical subgraph. Our Main Lemma (Corollary 2.12) then guarantees that we can choose a matching g_α from X_α into L_α which differs from j_α in only countable many edges, and which jilts the respective man which is the end vertex of the corresponding j-path starting from x, so that x has in fact the choice between $\varkappa$ men.

The most difficult problem is the proof that there are not more than $\varkappa$ elements of Z for which we have to make a match with men in $J = D(X)$; for in the case $|Z| > \varkappa$ the $\varkappa$ j-paths would be of no use. The proof of $|Z| \leq \varkappa$ will be guaranteed by a kind of Menger Theorem (Theorem 7.4). If we still notice the fact that the property of being popular in Γ is preserved in the subgraph $\Gamma(J,X)$ (Lemma 7.12), then we have prepared the essential tools to construct a matching from X into $D_\Gamma(X)$.

We now prove first the Menger-like Theorem 7.4. Then we formulate the stronger version 7.6 of Theorem 7.0 which will be proved by transfinite induction during the rest of the section. An important technical tool is Lemma 7.9, which shows at the same time why we could not restrict the definition of a popular woman to the existence of $\varkappa$ paths but required the condition on S of being stationary (there are several other reasons). In its proof we use for the first time the Choice Lemma of Chapter I which will be very useful in this section. Lemma 7.10 then shows $|Z| \leq \varkappa$, and the following lemmas yield the proof that X has the desired properties. We need some further definitions.

Definition. Let $G = (V,E)$ be a graph. If $T = (v_i : 0 \leq i \leq k)$ is a path in G, then let in(T) $= v_0$ and ter(T) $= v_k$ be the first and the last vertex of T. We denote the path $(v_i : 0 \leq i \leq j)$ by Tv_j and the path $(v_i : j \leq i \leq k)$ by v_jT for each $j \leq k$. $V(T) = \{v_i : 0 \leq i \leq k\}$ is the set of vertices of T. If $\mathcal{F}$ is a set of paths in G, then define $in[\mathcal{F}] = \{in(T) : T \in \mathcal{F}\}$, $ter[\mathcal{F}] = \{ter(T) : T \in \mathcal{F}\}$, and $V[\mathcal{F}] = \bigcup\{V(T) : T \in \mathcal{F}\}$. The same notions are defined for families of paths.

Lemma 7.3. Let $G = (V,E)$ be a graph (directed or undirected), let $\varkappa \geq \aleph_0$ be a regular cardinal, and let $\mathcal{F}$ a set of paths in G such that $|in[\mathcal{F}]| = \varkappa$. Then $\mathcal{F}$ has a subset of $\varkappa$ pairwise disjoint paths, or there exists a vertex $x \in V$ such that

$$d^-_{\mathcal{F}}(x) = |\{Tx : T \in \mathcal{F} \text{ and } x \neq in(T)\}| \geq \varkappa.$$

Proof. Let $\mathcal{H}$ be a $\subseteq$-maximal subset of pairwise disjoint paths of $\mathcal{F}$ and assume that $|\mathcal{H}| < \varkappa$. Take

$$\mathcal{G} = \{T \in \mathcal{F} : in(T) \notin V[\mathcal{H}]\}.$$

Especially we have $T \notin \mathcal{H}$ for every $T \in \mathcal{G}$. By the maximality of $\mathcal{H}$, the intersection $V(T) \cap V[\mathcal{H}]$ is not empty for each $T \in \mathcal{G}$. Since $|V[\mathcal{H}]| < \varkappa$

and $|in[\mathcal{F}]| = \varkappa$, we get $|in[\mathcal{G}]| = \varkappa$. Choose a path $T_y \subset \mathcal{G}$ with $in(T_y) = y$ for every $y \in in[\mathcal{G}]$, and let y' be the first vertex on T_y which is an element of $V(T_y) \cap V[\mathcal{H}]$. By the definition of $\mathcal{G}$, we have $y \neq y'$, further there exists an $x \in V[\mathcal{H}]$ such that $|\{y \in in[\mathcal{G}] : y' = x\}| = \varkappa$, since $|in[\mathcal{G}]| = \varkappa$. Consequently $|\{Tx : T \in \mathcal{G}$ and $x \neq in(T)\}| = \varkappa$ and $d^-_{\mathcal{F}}(x) \geq \varkappa$.

<u>Theorem 7.4</u>. Let $G = (V,E)$ be a graph, let λ be an infinite cardinal, and let C and D be disjoint subsets of V such that $|D| = \lambda$ and $|C| > |D|$. For each $c \in C$, let $\mathcal{F}_c$ be a c-disjoint set of paths in G from C to D. Then there exist a set $\mathcal{F}$ of pairwise disjoint paths from C to D and a $c_0 \in C$ such that $ter[\mathcal{F}_{c_0}] \subseteq ter[\mathcal{F}]$.

<u>Proof</u>. Without loss of generality we assume that every path of $\mathcal{F}_c$ starts in c. For each $c \in C$, define $\mathcal{F}'_c = \{T \subset \mathcal{F}_c : C \cap V(T) = \{c\}\}$ and $\mathcal{F}''_c = \{T \in \mathcal{F}_c : |C \cap V(T)| \geq 2\}$. Clearly $|\mathcal{F}_c| \leq \lambda$, since $|D| = \lambda$, and consequently $|\mathcal{F}'_c| \leq \lambda$ and $|\mathcal{F}''_c| \leq \lambda$.

Let S be the set of those $x \subset V \setminus C$ such that there exists an x-disjoint family of λ^+ paths from C to x.

We shall prove the following

<u>Claim</u>. There exists a $c_0 \in C$ such that $V(T) \cap S \neq \emptyset$ for all $T \subset \mathcal{F}'_{c_0}$.

First we show that this c_0 yields the desired family $\mathcal{F}$. Let $(T_\alpha : \alpha < \mu)$ be an enumeration of $\mathcal{F}'_{c_0}$ $(\mu \leq \lambda)$ and choose for each $\alpha < \mu$ an element $x_\alpha \in V(T_\alpha) \cap S$. Especially $x_\alpha \notin C$. Since $V(T_\alpha) \cap V(T_\beta) = \{c_0\}$, we have $x_\alpha \neq x_\beta$ for $\alpha \neq \beta$. We construct inductively disjoint paths T'_α from C to $ter(T_\alpha)$ which meet no path of $\mathcal{F}''_{c_0}$. Let T'_α be defined for every $\alpha < \nu$. We have $|\bigcup\{V(T'_\alpha) : \alpha < \nu\} \cup V[\mathcal{F}_{c_0}]| \leq \lambda$; but by the definition of S there exist λ^+ x_ν-disjoint paths from C to x_ν. Hence one of these paths, say $\overline{T}_\nu$, contains exactly one element of the set $\bigcup\{V(T'_\alpha) : \alpha < \nu\} \cup V[\mathcal{F}_{c_0}]$, namely x_ν. Take $T'_\nu = \overline{T}_\nu * x_\nu T_\nu$. This completes the definition of the sequence $(T'_\alpha : \alpha < \mu)$. For every $T \subset \mathcal{F}''_{c_0}$ there is a last vertex z_T on T which is an element of C. Since $z_T \neq c_0$, $\{z_T T : T \in \mathcal{F}''_{c_0}\}$ is a disjoint set of paths from C to D, and it is easy to see that $\mathcal{F} = \{T'_\alpha : \alpha < \mu\} \cup \{z_T T : T \subset \mathcal{F}''_{c_0}\}$ has the desired properties.

To prove the claim, assume that it is false and take $G' = G \setminus S$. Then, for each $c \in C$, the graph G' contains a path from c to D belonging to $\mathcal{F}'_c$. Choose such a path for each $c \in C$ and let $\mathcal{F}$ be the set of these paths. Clearly $|\mathcal{F}| \geq \lambda^+$, so we can assume that all the paths in $\mathcal{F}$ have the same length n for some $n \in \omega$, $n \neq 0$. Since $|D| = \lambda$, $\mathcal{F}$ cannot have a disjoint subset of cardinality λ^+. So Lemma 7.3 yields a $y \in V \setminus S$ such that $d^-_{\mathcal{F}}(y) \geq \lambda^+$. We have $y \notin C$, since only the initial vertex of any path in $\mathcal{F}$ belongs to C. Let $\overline{\mathcal{F}} = \{Ty : T \in \mathcal{F}$ and $y \neq in(T)\}$ and $\mathcal{F}_1 = \{T : T*(y) \in \overline{\mathcal{F}}\}$. $\overline{\mathcal{F}}$ is a family of paths from C to y and $|\overline{\mathcal{F}}| \geq \lambda^+$, therefore $|\mathcal{F}_1| \geq \lambda^+$, since otherwise we could find in $\mathcal{F}$ λ^+ edges from C to y, so $y \in S$ would yield a contradiction. By the same reason, $\mathcal{F}_1$ cannot have a subset of λ^+ pairwise disjoint paths. Consequently Lemma 7.3 again yields a $y_1 \in V \setminus S$ such that $d^-_{\mathcal{F}_1}(y_1) \geq \lambda^+$. The iteration of this argument yields distinct vertices $y, y_1, \dots, y_{n+1}$ and λ^+ paths in $\mathcal{F}$ each of which contains $y, y_1, \dots, y_{n+1}$, a contradiction to the fact that every path in $\mathcal{F}$ is of length n. This proves the claim and completes the proof of the theorem.

We now define the critical subgraph $\Delta(\Gamma)$. To be more precise, if $\varkappa > 1$ and $\overline{\Sigma} = (\Sigma_\alpha : \alpha < \varkappa)$ is an obstructive ladder for $\Sigma = \vee\overline{\Sigma}$, we define, by induction on $\varkappa$, a critical graph $\Delta(\overline{\Sigma})$ of Σ and a marriage $f(\overline{\Sigma})$ of $\Delta(\overline{\Sigma})$ in dependence of $\overline{\Sigma}$ and write $\Delta(\Sigma)$, $f(\Sigma)$, if $\overline{\Sigma}$ is fixed.

If Σ is a 1-obstruction, we choose $\Delta(\Sigma)$ as in Lemma 5.4 and $f(\Sigma)$ as a marriage of $\Delta(\Sigma)$. Now let $\varkappa \in Rc \setminus \{1\}$, $\overline{\Sigma} = (\Sigma_\alpha : \alpha < \varkappa)$ an obstructive ladder for $\Sigma = \vee\overline{\Sigma}$, and $S = \{\alpha < \varkappa : \Sigma_\alpha$ is a ρ_α-obstruction for some $\rho_\alpha < \varkappa\}$. For $\alpha \in S$ let $\overline{\Sigma}_\alpha$ be Σ_α, if $\rho_\alpha = 1$, and a fixed obstructive ladder for Σ_α, if $\rho_\alpha > 1$. So, by induction, $\Delta(\overline{\Sigma}_\alpha)$ and $f(\overline{\Sigma}_\alpha)$ are defined, and we take

$$\Delta(\overline{\Sigma}) = \vee\{\Delta(\overline{\Sigma}_\alpha) : \alpha < \varkappa\} \text{ and}$$

$$f(\overline{\Sigma}) = \cup\{f(\overline{\Sigma}_\alpha) : \alpha < \varkappa\}.$$

Lemma 7.5.

If we take the notation of the preceeding definitions, then

(a) $f(\overline{\Sigma})$ is a marriage of $\Delta(\overline{\Sigma})$.

(b) For every $\alpha \in S$ we have $f(\overline{\Sigma}_\alpha) = f(\overline{\Sigma}) \restriction M_{\Sigma_\alpha}$.

(c) $\Delta(\overline{\Sigma})$ is critical.

(d) $|W_\Sigma \setminus W_{\Delta(\overline{\Sigma})}| \leq \varkappa$.

Proof. (a) and (b) follow directly from the definition of $f(\overline{\Sigma})$ and $\Delta(\overline{\Sigma})$, (c) and (d) are proved by induction on $\varkappa$ - see also Lemma 5.4.

We can now formulate the stronger version of Theorem 7.0.

Theorem 7.6.
Let $\varkappa = 1$ or $\varkappa$ be a regular uncountable cardinal, let $\Gamma = (M,W,E)$ be a $\varkappa$-obstruction, and let $\overline{\Lambda}$ be an obstructive $\varkappa$-ladder for Γ if $\varkappa \neq 1$. If $f = f(\overline{\Lambda})$ is chosen according to Lemma 7.5, then there exist a subset X of W and a matching h of X into $D_\Gamma(X)$ such that

(a) $D_\Gamma(X)$ is unmatchable and

(b) $|h \setminus f| \leq \varkappa$.

The rest of this section is devoted to the proof of the theorem.

If $\varkappa = 1$, then it is obvious that $X = W$ and $h = f$ have the desired properties. Therefore we always assume that $\varkappa$ is regular and uncountable. The prerequisite for the application of Corollary 2.12 to get enough unmarried men for our popular women is the following lemma.

Lemma 7.7.
If Theorem 7.6 is true for Γ, then it is possible to choose X in such a way that additionally to (a) and (b) the following property holds:

(c) The greatest critical subgraph of $\overleftarrow{\Gamma}(X,D(X))$ is $(\emptyset,\emptyset,\emptyset)$.

Proof. Remember that $\overleftarrow{\Gamma}$ is (W,M,E). Let X and f have the properties of Theorem 7.6, and let Σ be the greatest critical subgraph of $\overleftarrow{\Gamma}(X,D(X))$ (which exists by Theorem 2.4). Take $X' = X \setminus M_\Sigma$. Since Σ is saturated in $\overleftarrow{\Gamma}(X,D(X))$, the restriction of the matching h of X to X' is a matching into $D_\Gamma(X')$: If $a \in X'$ and $m = h(a)$, then of course we have $E<m> \subseteq X$; further $m \notin W_\Sigma$ ($\subseteq D(X)$!) since otherwise $h \restriction M_\Sigma$ would not match m which contradicts the fact that Σ is critical. Consequently $E<m> \subseteq X'$, since Σ is saturated. If g would be a matching of $D(X')$ into X', then $g \cup h \restriction W_\Sigma$ would be a matching of $D(X)$, since $h[M_\Sigma] = W_\Sigma$, and this would be a contradiction to property (a). The greatest critical subgraph of $\overleftarrow{\Gamma}(X',D(X'))$ is $(\emptyset,\emptyset,\emptyset)$, and our assumption is true if we replace X and h by X' and $h \restriction X'$.

Now let $\overline{\Lambda} = (\Lambda_\alpha : \alpha < \varkappa)$, $\Lambda_\alpha = (M_\alpha, W_\alpha, E_\alpha)$ and $f_\alpha = f \restriction M_\alpha$. As above define $S = \{\alpha < \varkappa : \Lambda_\alpha$ is a ρ-obstruction for some $\rho < \varkappa\}$ and $\overline{S} = \varkappa \setminus S$. Let $\overline{\Pi}$ be the tower corresponding to $\overline{\Lambda}$, i.e. for all $\alpha \leq \varkappa$ we have $\Pi_\alpha = \bigvee\{\Lambda_\beta : \beta < \alpha\}$, and each Π_α is saturated in Γ.

We assume that Theorem 7.6 is true for every ρ-obstruction such that $\rho<\kappa$. Since $f(\overline{\Lambda}_\alpha) = f_\alpha$ for some fixed ρ-obstruction $\overline{\Lambda}_\alpha$ of Λ_α by Lemma 7.5, we get by the hypothesis of the induction for each $\alpha\in S$ a set $X_\alpha \subseteq W_\alpha$ and a matching h_α of X_α into $\underline{L_\alpha} := D_{\Lambda_\alpha}(X_\alpha)$ such that

(a) $D_{\Lambda_\alpha}(X_\alpha)$ is unmatchable in Λ_α,

(b) $|h_\alpha \setminus f_\alpha| \leq \rho_\alpha < \kappa$, and

(c) The greatest critical subgraph of $\overleftarrow{\Gamma}(X_\alpha, D_{\Lambda_\alpha}(X_\alpha))$ is $(\emptyset,\emptyset,\emptyset)$.

We now give once more the definition of the function j which we discussed at the begin of this section:

$$j_\alpha = h_\alpha \cup (f_\alpha \cap (M_\alpha \setminus h_\alpha[X_\alpha]) \otimes (W_\alpha \setminus X_\alpha))$$

for every $\alpha\in S$ and

$$j = \bigcup\{j_\alpha : \alpha\in S\}.$$

Obviously we have

(b') $|j_\alpha \setminus f_\alpha| = |j_\alpha \setminus f| \leq \rho_\alpha < \kappa$ for every $\alpha \in S$.

Lemma 7.8.

(1) $|W_\alpha \setminus j[M_\alpha]| \leq \rho_\alpha < \kappa$ for each $\alpha<\kappa$.

(2) $|W \setminus j[M]| \leq \kappa$

Proof. Clearly (1) implies (2). Let $\alpha\in S$. By Lemma 7.5, $f \restriction M_\alpha = f(\overline{\Lambda}_\alpha) = f_\alpha$, and if $\Delta(\Gamma)$ is the critical subgraph married by f, then further $\Delta(\overline{\Lambda}_\alpha) = \Delta(\Gamma)\cap\Lambda_\alpha$ and $|W_\alpha \setminus W_{\Delta(\overline{\Lambda}_\alpha)}| \leq \rho_\alpha$. If we regard $w \in W_\alpha$ which is not contained in an edge of j_α, then there are two possibilities: It is not contained in an edge of f_α or it is contained in an edge of f_α which is not compatible with h_α.

The first case can only occur ρ_α times, since $f_\alpha[M_\alpha] = W_{\Delta(\overline{\Lambda}_\alpha)}$ and $|W_\alpha \setminus W_{\Sigma(\overline{\Lambda}_\alpha)}| \leq \rho_\alpha$, and the second can occur only ρ_α times, since $|j_\alpha \setminus f_\alpha| \leq \rho_\alpha$. So we get (1).

The definition of a j-path, of a stationary family of paths and so on remains the same as on page 85. In the same way we ask the reader to remember the definition of the sets U_i, N_i, R_i, and Z_i $(i\in\omega)$, and of U, N, R, Z, J, and X on page 86.

The following lemma is a variation of Fodor's Theorem with help of our notions and will be an important technical tool for the following considerations:

Lemma 7.9. If $k<\omega$, then there does not exist a disjoint stationary family $\mathcal{T}$ of j-paths such that every member of $\mathcal{T}$ starts in N_k.

Proof. To get a contradiction, we choose k as the least $n\in\omega$ such that our assumption is false for n. Let $\mathcal{T} = (T_\alpha : \alpha\in S_1)$ be a stationary disjoint family of j-paths starting in N_k such that S_1 is stationary in $\varkappa$, $S_1 \subseteq S$, and $ter(T_\alpha)\in L_\alpha$ for every $\alpha\in S_1$.

Case 1. $k = 0$.

For each $\alpha\in S_1$ choose $\varphi(\alpha)$ as the index of the rung of the ladder $\overline{\Lambda}$ from which the path T_α starts. Since every path of the family $\mathcal{T}$ starts in $N_0 = W\ j[M]$, we get $\varphi(\alpha)<\alpha$ for each $\alpha\in S_1$ by Lemma 7.1. By Fodor's Theorem, the function φ is constant on a stationary subset S_2 of S_1 - let β be the value of φ on S_2. Since $|S_2| = \varkappa$ and $\mathcal{T}$ is disjoint, we have $|N_0\cap W_\beta| = \varkappa$. On the other hand, as in the proof of Lemma 7.8, for any $m\in N_0\cap W_\beta$ there are only two possibilities: Either m is contained in no edge of f, and by the construction of f this case can occur only less than $\varkappa$ times, or m belongs to an edge of f which is incompatible with h_α, and this case occurs less than $\varkappa$ times by the inductive hypothesis. So we get for $k = 0$ the desired contradiction.

Case 2. $k>0$.

By the definition of N_k, we have $N_k\subseteq j[M]$. Let $w_\alpha = in(T_\alpha)$ and $a_\alpha = j(w_\alpha)$ for each $\alpha\in S_1$. Since $U_k = E[R_{k-1}]$ and $N_k = j[U_k]$, we get $a_\alpha\in E[R_{k-1}]$. Now regard, for every $x\in R_{k-1}$, the set $O(x) = \{\alpha\in S_1 : a_\alpha\in E\langle x\rangle\}$. We show that these sets are non-stationary and suppose, to get a contradiction, that there exists some $x\in R_{k-1}$ such that $O(x)$ is stationary. Since $\mathcal{T}$ is disjoint, there is at most one path in $\mathcal{T}$ which contains x. This is unessential for the property of being stationary, therefore we omit this path, if necessary, and assume that no path in $\mathcal{T}$ contains x.

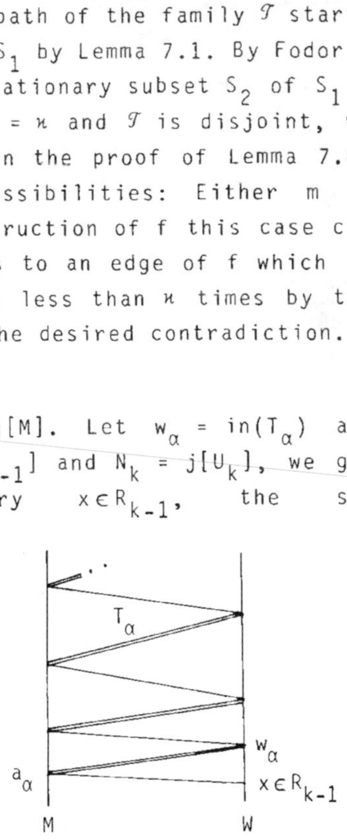

Now it could happen that some a_α's are endvertices of paths in $\mathcal{T}$. If this would be the case for each α in a stationary subset $O_1(x)$ of $O(x)$, then $((x,a_\alpha) : \alpha\in O_1(x))$ would be an x-disjoint stationary family of j-paths, since each path in $\mathcal{T}$ ends in some L_α, and this would contradict the fact that x, as an element of $R_{k-1} = N_{k-1}\cap Q$, is not popular. So $O_2(x) = O(x)\setminus O_1(x)$ is stationary, and again $((x,a_\alpha,w_\alpha) * T_\alpha : \alpha\in O_2(x))$ is an x-disjoint stationary family of j-paths, which contradicts $x \notin P$. So we have shown that $O(x)$ is not stationary, for each $x \in R_{k-1}$. On the other hand, the set $S_1 = \cup\{O(x) : x \in R_{k-1}\}$ is stationary, especially $|R_{k-1}| = \kappa$ by I.6.7. By the Choice Lemma I.6.9, we can find a stationary subset S_3 of S_1 such that each $\alpha \in S_3$ belongs to some $O(x_\alpha)$ with $a_\alpha \in E\langle x_\alpha\rangle$. Clearly x_α cannot be an endvertex by definition of a j-path. Define, for $\alpha \in S_3$, T'_α to be the path $x_\alpha T_\alpha$ if x_α lies on T_α, and to be the path $(x_\alpha, a_\alpha, w_\alpha) * T_\alpha$ otherwise. Let $G' = (V', E')$ be the graph whose vertices are those vertices of Γ which are lying on some T'_α for some $\alpha \in S_3$ and whose edges are the edges belonging to such a path. Since $\mathcal{T}$ is disjoint, each vertex $v \in V'$ lies on at most two paths of the family $\mathcal{T}' = (T'_\alpha : \alpha \in S_3)$. If $v \in V'$ is a fixed vertex and $C_n(v)$ is the set of all vertices of G' which are endvertices of some path in G' of length n which starts in v, then it is easy to see by complete induction that each $C_n(v)$ is finite. So every connected component C of G' is countable; especially the set $\{\alpha \in S_3 : V(T'_\alpha) \subseteq C\}$ is countable and all the more not stationary. But S_3 is stationary, and by the Choice Lemma again we can find a stationary subset S_4 of S_3 such that, for any $\alpha, \beta \in S_4$ with $\alpha \neq \beta$, the paths T'_α and T'_β are belonging to different connected components of G'. So $\mathcal{T}'' = (T'_\alpha : \alpha \in S_4)$ is a stationary disjoint family of j-paths which are starting from N_{k-1}, and this contradicts the choice of k.

<u>Lemma 7.10</u>. $|Z_k| \leq \kappa$ for every $k \in \omega$.

<u>Proof</u>. Suppose that there exists a $k \in \omega$ such that $|Z_k| > \kappa$. Since every woman in Z_k is popular, we get for each $x \in Z_k$ an x-disjoint stationary family $\mathcal{T}(x)$ of j-paths. Roughly speaking we have reached the situation of Theorem 7.4 with $C = Z_k$ and $D = S$, if we admit the elements $\alpha \in S$ which have the property that the endvertex of some path T_β starting in Z_k lies in L_α, as new endvertices of paths. Therefore let $G = (V, E_G)$ be the following <u>directed</u> graph: $V = M \cup W \cup S$ and $E_G = \{(a, j(a)) : a \in M\} \cup \{(w,m) : \{w,m\} \subset E\} \cup \{(m,\alpha) : \alpha \in S \text{ and } m \in L_\alpha\}$.

A path in G which starts at some women necessarily is, after deleting a possible edge (m,α), a j-path in Γ, since $W \cap M = \emptyset$. It is easy to see that the sets $C = Z_k$, $D = S$, and $\mathcal{F}_x = (T * (ter(T), \alpha) : T \in \mathcal{T}(x) \text{ and } ter(T) \in L_\alpha)$ are fulfilling the assumptions of Theorem 7.4. So we get a disjoint family $\mathcal{T}'$ of paths in G from Z_k to S such that $ter[\mathcal{F}_{x_0}] \subseteq ter(\mathcal{T}')$ for some $x_0 \in Z_k$. Let $\mathcal{T}$ be the family which results from $\mathcal{T}'$ by deleting the last edge of each path of $\mathcal{T}'$. Then $\mathcal{T}$ is a disjoint family of j-paths in Γ, which is stationary since $ter[\mathcal{T}_{x_0}]$ is stationary, and whose paths are starting in $Z_k \subseteq N_k$. This contradicts Lemma 7.9, and the lemma is proved.

Lemma 7.11. If $S' = \{\alpha \in S : L_\alpha \cap U = \emptyset\}$, then $S \setminus S'$ is not stationary in $\varkappa$.

Proof. Let us suppose that $D = S \setminus S'$ is stationary in $\varkappa$. Since $\varkappa > \aleph_0$, there must exist a $k \in \omega$ such that the set $D_k = \{\alpha \in S : L_\alpha \cap U_k \neq \emptyset\}$ is stationary in $\varkappa$ (see Lemma I.6.7). Clearly $k \neq 0$, since $U_0 = \emptyset$.

For $x \in R_{k-1}$ define $O(x) = \{\alpha \in S : E\langle x\rangle \cap L_\alpha \neq \emptyset\}$. Since x is not popular, the set $O(x)$ is not stationary. On the other hand D_k is the union of the sets $O(x)$, $x \in R_{k-1}$. By the Choice Lemma, there exist a stationary subset S_1 of D_k, distinct vertices $x_\alpha \in R_{k-1}$, and distinct vertices $a_\alpha \in L_\alpha$ such that $a_\alpha \in E\langle x_\alpha\rangle$ for each $\alpha \in S_1$. The family $\mathcal{T} = ((x_\alpha, a_\alpha) : \alpha \in S_1)$ is stationary and consists of j-paths which are starting in $R_{k-1} \subseteq N_{k-1}$. This contradiction to Lemma 7.9 shows that $S \setminus S'$ is not stationary in $\varkappa$.

Lemma 7.12. If $x \in X$ is popular, then there exists an x-disjoint stationary family of j-paths in the graph $\Gamma(J,X)$, i.e. x is popular in $\Gamma(J,X)$.

Proof. Let $\mathcal{T} = (T_\alpha : \alpha \in S_1)$ be an x-disjoint family of j-paths which makes $x \in X$ popular in Γ; that means that S_1 is a stationary subset of S and $ter(T_\alpha) \subset L_\alpha$ for each $\alpha \in S_1$.

Take $D = \{\alpha \in S_1 : V(T_\alpha) \cap U \neq \emptyset\}$. We show that D is not stationary in $\varkappa$. Then the family $\mathcal{T}' = (T_\alpha : \alpha \in S_1 \setminus D)$ is an x-disjoint stationary family of j-paths whose members do not contain any vertex which belongs to U, and consequently they cannot contain any vertex which belongs to $R = \cup\{R_i ; i \in \omega\}$ since $U_k = E[R_{k-1}]$ for each $k \geq 1$. So $\mathcal{T}'$ has the desired properties.

Suppose for contradiction that D is stationary in $\varkappa$. By Lemma I.6.7, there exists a $k \in \omega$ such that $D_k = \{\alpha \in S_1 : V(T_\alpha) \cap U_k \neq \emptyset\}$ is stationary in $\varkappa$. Lemma 7.11 shows that the set of $\alpha \in S$ for which T_α ends in U_k cannot be stationary. Therefore, choosing $a_\alpha \in V(T_\alpha) \cap U_k$ for each $\alpha \in D_k$, the set $D_k' = \{\alpha \in D_k : a_\alpha$ is not the endvertex of $T_\alpha\}$ is stationary, and the family $(j(a_\alpha)T_\alpha : \alpha \in D_k')$ is a disjoint stationary family of j-paths which are starting in $j[U_k] = N_k$. This contradicts Lemma 7.9. Consequently D is not stationary in $\varkappa$, and the lemma is proved.

<u>Lemma 7.13</u>. $D_\Gamma(X)$ is unmatchable

<u>Proof</u>. To get a contradiction we suppose that g is a matching of $J = D_\Gamma(X)$ in Γ. Clearly $g[J] \subseteq X$. By Lemma 7.11, we have $L_\beta \subseteq J$ for each element β of the stationary set $S_1 = \{\alpha \in S : L_\alpha \cap U = \emptyset\}$. Since $L_\beta = D_{\Lambda_\beta}(X_\beta)$ is unmatchable in Λ_β by the hypothesis of the induction, and since there are no edges from L_β to some W_α with $\beta < \alpha$, there must exist some $m_\beta \in L_\beta$ and some $\varphi(\beta) < \beta$ such that $g(m_\beta) \in W_{\varphi(\beta)}$. Fodor's Theorem yields a stationary subset S_2 of S_1 and a $\gamma < \varkappa$ such that φ has the constant value γ on S_2. Take $\delta = \gamma + 1$ and $\Pi_\delta = \vee\{\Lambda_\beta : \beta < \delta\}$. We construct a stationary disjoint family of j-paths starting in N, which will contradict Lemma 7.9 since $\varkappa$ is uncountable. The paths are defined "from above" with help of the matchings f and g. For every $\beta \in S_2$ we define the sequence $p_\beta = (m_0^\beta, w_0^\beta, m_1^\beta, w_1^\beta, \dots)$ as follows: $m_0^\beta = m_\beta$, $m_{n+1}^\beta = f(w_n^\beta)$, $w_n^\beta = g(m_n^\beta)$, if a corresponding edge exists in Γ; otherwise the definition terminates. This sequence cannot be infinite: If this would be the case, take

$$f' = (f \setminus \{\{m_{n+1}^\beta, w_n^\beta\} : n \geq 0\}) \cup \{\{m_n^\beta, w_n^\beta\} : n \geq 1\}.$$

f' is a marriage of the critical subgraph $\Delta(\Gamma)$ and $w_0^\beta \in W_\Delta \setminus f'[M_\Delta]$, a contradiction. So there exists a last woman in the sequence, which we call w_β. If $n+1$ is the number of women in the sequence p_β, we get by $(w_\beta, m_n^\beta, w_{n-1}^\beta, \dots, w_0^\beta, m_\beta)$ an f-alternating path T_β from w_β to $m_\beta \in L_\beta$, for any $\beta \in S_2$.

<u>Claim</u>. If D is the set of $\beta \in S_2$ for which $w_\beta \notin j[M]$ or $f(w_\beta) \neq j(w_\beta)$ or T_β is not a j-path, then $|D| < \varkappa$.

If the claim is proved, $\mathcal{T} = (T_\beta : \beta \in S_2 \setminus D)$ is a stationary disjoint family of j-paths. Since $w_\beta \in j[M]$ for $\beta \in S_2 \setminus D$, the construction of p_β must have terminated because $f(w_\beta) = j(w_\beta) \notin J$. This implies that $j(w_\beta) \in U_k$ for some $k < \omega$ and so $w_\beta \in N_k$. So $\mathcal{T}$ is a stationary disjoint family of j-paths starting from N, contradicting 7.9.

Since $|W_\alpha \setminus j[M_\alpha]| < \varkappa$ by Lemma 7.8, for each $\alpha < \varkappa$, it suffices for the proof of the claim to show that $|f \upharpoonright \Pi_\delta \setminus j| < \varkappa$. Consider an edge e of $f \upharpoonright \Pi_\delta \setminus j$. Then $e \in f_\alpha \setminus j_\alpha$ for some $\alpha < \delta$ by the choice of f and j. By the definition of j_α, $e \in M_\alpha \otimes X_\alpha$, since there exists no f_α-edge in $L_\alpha \otimes (W_\alpha \setminus X_\alpha)$. But X_α is matched by h_α and $|h_\alpha \setminus f_\alpha| < \varkappa$ by condition (b), so $|f_\alpha \setminus j_\alpha| < \varkappa$ and $|f \upharpoonright \Pi_\delta \setminus j| = |\bigcup\{f_\alpha \setminus j_\alpha : \alpha < \delta\}| < \varkappa$. This proves the claim and completes the proof of Lemma 7.13.

<u>Lemma 7.14</u>. There exists a matching h of X into J such that $|h \setminus f| \leq \varkappa$.

<u>Proof</u>. Let $|Z| = \mu \leq \varkappa$ (Lemma 7.10), and let $(s_\alpha : \alpha < \mu)$ be an enumeration of Z. By transfinite induction, we construct a sequence $(T_\alpha : \alpha < \sigma)$ of pairwise disjoint j-paths, $\sigma \leq \mu$, such that $V(T_\alpha) \subseteq J \cup X$ and T_α is a j-path from Z to $L_{\varphi(\alpha)}$ for some $\varphi(\alpha) \subset \{\beta : L_\beta \cap U = \emptyset\}$; further $\varphi(\alpha_1) \neq \varphi(\alpha_2)$ for $\alpha_1 \neq \alpha_2$, and there exists a matching g_α of $X_{\varphi(\alpha)}$ into $L_{\varphi(\alpha)}$ such that $|g_\alpha \setminus j| \leq \aleph_0$ and $ter(T_\alpha) \notin g_\alpha[X_{\varphi(\alpha)}]$.

We have $Z \subseteq P \cap X$, hence Lemma 7.12 yields a $\varphi(0)$ such that $L_{\varphi(0)} \subseteq J$, and a j-path T_0 from s_0 to $L_{\varphi(0)}$ such that $V(T_0) \subseteq J \cup X$. We can assume that T_0 and $L_{\varphi(0)}$ only have the endvertex $ter(T_0)$ in common - otherwise we take the path T_0 up to the first vertex in $L_{\varphi(0)}$ and delete the following edges. Since the graph $\overleftarrow{\Gamma}(X_{\varphi(0)}, L_{\varphi(0)})$ has no critical subgraph except $(\emptyset, \emptyset, \emptyset)$ by property (c), and has the marriage $j_{\varphi(0)}$, there exists, by Corollary 2.12, a matching $\bar{g}_0$ of $X_{\varphi(0)}$ into $L_{\varphi(0)}$ such that $ter(T_0) \notin \bar{g}_0[X_{\varphi(0)}]$. Let $m_0 = ter(T_0)$ and define by recursion $m_{n+1} = \bar{g}_0(w_n)$ and $w_n = j_{\varphi(0)}(m_n)$, if m_n belongs to an edge of $j_{\varphi(0)}$; otherwise the definition terminates. Now g_0 results from $j_{\varphi(0)}$ by replacing the edges $\{m_n, w_n\}$, $n < k \leq \omega$, by the edges $\{w_n, m_{n+1}\}$, $n < k \leq \omega$. Clearly g_0 is a matching of $X_{\varphi(0)}$ into $L_{\varphi(0)}$ with the properties $ter(T_0) = m_0 \notin g_0[X_{\varphi(0)}]$ and $|g_0 \setminus j_{\varphi(0)}| = |g_0 \setminus j| \leq \aleph_0$. Now let, for each $\beta < \alpha < \mu$, T_β, g_β, and $\varphi(\beta)$ be defined with the properties mentioned above. Take $V_\alpha = \bigcup\{V(T_\beta) \cup (g_\beta \setminus j)[M \cup W] : \beta < \alpha\}$. Clearly $|V_\alpha| \leq \aleph_0 \cdot |\alpha| < \varkappa$. If $Z \subseteq V_\alpha$, we are done and choose $\sigma = \alpha$. Otherwise let $\gamma = \gamma(\alpha)$ be the first ordinal such that $s_\gamma \notin V_\alpha$. As in the first step of the construction it follows from Lemma 7.12 and $s_\gamma \in X$ that we can find a j-path T_α in $\Gamma(J,X)$ which has s_γ as its first vertex and ends in some $L_{\varphi(\alpha)}$ such that $L_{\varphi(\alpha)} \subseteq J$ and T_α and $L_{\varphi(\alpha)}$ have only the vertex $ter(T_\alpha)$ in common. And the property that s_γ is popular guarantees that we can choose T_α in such a way that T_α does not meet V_α ($|V_\alpha| < \varkappa$!) and $\varphi(\alpha) > \sup\{\varphi(\beta) : \beta < \alpha\}$. Finally we construct g_α in complete analogy to the first step; so g_α is a matching of $X_{\varphi(\alpha)}$ into $L_{\varphi(\alpha)}$ such that $|g_\alpha \setminus j| \leq \aleph_0$ and $ter(T_\alpha) \notin g_\alpha[X_{\varphi(\alpha)}]$.

Now we are able to match the elements of Z. Observe that $Z \subseteq \bigcup\{V_\alpha : \alpha < \sigma\} = Z_1$. If $T_\alpha = (w_0^\alpha, m_1^\alpha, \ldots, w_{n(\alpha)-1}^\alpha, m_{n(\alpha)}^\alpha)$, we define $h(w_k^\alpha) = m_{k+1}^\alpha$ for any $k < n(\alpha)$, as we motivated at the begin of this section. If $x \in Z_1 \cap X$ does not belong to any T_α, it belongs to an edge of $g_\alpha \setminus j$ for some $\alpha < \sigma$. Define $h(x) = g_\alpha(x)$. In both cases h matches into $D(X) = J$, since $V(T_\alpha) \subseteq J \cup X$ and $L_{\varphi(\alpha)} \subseteq J$ for every $\alpha < \sigma$.

If $x \in X$ is not an element of Z_1, then especially $x \in X \setminus Z$, and therefore x must belong to any edge of j, and $j(x) \in J$ by the definition of X and J. Take $h(x) = j(x)$. This completes the definition of h, and our construction shows that h is a matching of X into $D_\Gamma(X)$. Further $|h \setminus f| \leq |Z_1| + |j \setminus f| \leq \kappa \cdot |\sigma| + \kappa = \kappa$. This completes the proof of Lemma 7.14.

Now Lemma 7.13 and Lemma 7.14 imply Theorem 7.6.

Chapter III

STRUCTURAL PROPERTIES OF FAMILIES

In the literature one can find several necessary and sufficient conditions for the existence of a marriage of a countable family of sets. One purpose of this chapter is to present these conditions. We now formulate them though the reader probably will understand them in detail not before the study of the corresponding section.

The first condition was presented by Nash-Williams in [N1]. He defined, by transfinite recursion, so-called margin functions m_α for every ordinal α and conjectured (Corollary 8.2) that a countable family has a marriage if and only if

(a) $m_\alpha(X) \geq 0$ for every set X of women and every ordinal α.

This was proved by Damerell and Milner in [DM]. In [N2] Nash-Williams gives several variations of the notion of a margin function. We are going to present a further variation in Section 8 and will show that m_α can be defined as follows: We define, for any family $F = (F(i) : i \in I)$, the set

$$J_\alpha := \{i : \text{There is a } \beta < \alpha \text{ with } m_\beta(F(i)) \in \omega\},$$

and we say that <u>B can be left unmarried at stage α</u>, if $|\{i \notin J_\beta : F(i) \subseteq Z\}| + |B| \leq m_\beta(Z)$ for each $\beta < \alpha$ and each $Z \supseteq B$ with $m_\beta(Z) \in \omega$ [1]. Then take $m_\alpha(X) := \sup\{|B| + 1 : B \subset\subset X \text{ can be left unmarried at stage } \alpha\} - 1$.

Clearly the following condition is necessary for the existence of a marriage of F.

(b) If $F \restriction J$ has a marriage and $i \in \operatorname{dom} F$, then $F \restriction J \cup \{i\}$ has a marriage.

In Section 3 we prove that a countable family has a marriage if (b) holds for each J and i.

In [St1] the following condition was introduced:

(c) If K is critical in F and $F(i) \subseteq F(K)$, then $i \in K$.

It was shown in [PS1] that a countable family has a marriage if and only if (c) holds. This condition is investigated in Section 3, too.

---

[1] In this chapter, the letters J,K,L,M always denote sets of men, and A,B,C,D,X,Y,Z always denote sets of women. Similarly i,j denote men and a,b denote women.

A fourth criterion was given in [HPS2], and we present a similar criterion in Section 8. There we define so-called demand functions J_α: If X is a set of women, we put $J_\alpha(X) := \{i \in dom(F) : i$ demands a woman of X at stage β for some $\beta < \alpha\}$. Then we say that i demands a woman of X at stage α if $i \notin J_\alpha(F(dom\,F))$ and there is a set $B \subset\subset X \cap F(i)$ such that, for each $Y \supseteq B$ and each $a \in F(i)$, there is a finite set D with $a \in D$ and $|J_\alpha(D) \setminus J_\alpha(Y)| \geq |D \setminus Y|$.

We shall show that a countable family has a marriage if and only if

(d) $|J_\alpha(B)| \leq |B|$ for every ordinal α and every finite set B.

A further criterion was given by Nash-Williams in [N3]. He assigned to every sequence k of women of a family F a value $p_F(k) \in \{-\infty\} \cup \mathbb{Z} \cup \{\infty\}$ and proved that any countable family F has a marriage if and only if

(e) $p_F(k) \geq 0$ for every k.

A variation of this criterion is proved in Section 5.

In Section 2 we define the notion of a frame (in a family F). Let H be a family such that $dom\,H \subseteq dom\,F$ and $H(i) \subset\subset F(i)$ for every $i \in dom\,H$, and let $r : dom\,H \to On$ be a function. H is called <u>a frame in F</u> if, for all $i \in dom\,H$, $a \in F(i)$, and $L \subseteq dom\,H$ with $i \in L$, there is a $K \subset\subset \{j : r(j) < r(i)\}$ such that $a \in H(K \cup L)$ and $|H(K) \setminus H(L)| \leq |K \setminus L|$.

Frames are quite useful to extend results about Hall families to other families. We are going to show in Section 2 that a countable family has a marriage if and only if

(f) Every frame in F fulfils Hall's condition.

We will show that the conditions (a) - (f) are equivalent. For this reason we define a set I_F as follows:

$$I_F := \{i : \text{There are a finite set A and a set K critical in } F \setminus A \text{ such that } F(i) \subseteq F(K) \cup A\}.$$

The family $F \restriction I_F$ was investigated in [HPS1], and the results of this paper can be found in Sections 6 and 7. Let (*) be one of the conditions (a) - (f). Then we shall prove that (*) holds for F if and only if $F \restriction I_F$ has a marriage.

We have already seen the importance of critical families in Chapter II. In this chapter we shall give several characterizations of critical families. In Section 2 we show that a family is critical if and only if there is a critical frame H in F with dom H = dom F. This shows that critical families almost are Hall families. In Section 4 we give three inductive definitions of the class of all critical families. Further Aharoni used in [A1] the function p_F to characterize critical families. This characterization can be found in Section 5. Finally, in Section 8, we describe critical families with help of the demand sets $J_\alpha(X)$ and with help of the margin functions $m_\alpha(X)$.

But first, in Section 1, we are going to investigate Hall families.

§1. Hall families

A family $F = (F(i) : i \in I)$ is called a Hall family if $F(i)$ is finite for each $i \in I$. P.Hall was the first to investigate the question if a finite Hall family possesses a marriage. Let us sketch one of his motivations. Van der Waerden recognized that the following problem of group theory is of pure set theoretic nature ([vdW]):

> If G is a finite group and if U is a subgroup of G, do the left cosets and the right cosets of U in G possess a common set of representatives?

Let G and U be fixed and choose a set A of representatives of the left cosets and a set B of representatives of the right cosets of U in G, also called a left transversal of U in G and a right transversal of U in G. Let $H = (H(a) : a \in A)$ be defined by $H(a) = \{b \in B : aU \cap Ub \neq \emptyset\}$.

Claim. If H has a marriage, then there exists a subset C of G which is a right and a left transversal of U in G.

To see this let f be a marriage of H. Since $|A| \cdot |U| = |G| = |B|\,|U|$ and G is finite, $f : A \to B$ must be bijective. Now, by definition, $aU \cap Uf(a) \neq \emptyset$ for each $a \in A$. Choose, for each $a \in A$, an element c_a of $aU \cap Uf(a)$ and take $C = \{c_a : a \in A\}$. Then C has the desired properties.

We have seen that the group theoretic problem is reduced to the question whether H has a marriage. If an arbitrary family F possesses a marriage, it satisfies the following condition which we call Hall's condition:

$|J| \leq |(F(J)|$ for each finite subset J of I.

P.Hall showed that for finite families this condition is also sufficient for the existence of a marriage (see Theorem 1.2). This implies that $H \in \mathcal{M}$, which had already been proved by v.d.Waerden and D.König.

Claim. H has a marriage.

To prove the claim, we show that H satisfies Hall's condition. Let $J \subseteq A$, $a \in J$, and $x \in aU$. Then $aU = xU$, $xU = Ub$ for some $b \in B$, and $x \in aU \cap Ub$. So $\bigcup\{aU : a \in J\} \subseteq \bigcup\{Ub : b \in H(J)\}$, and consequently $|J| \cdot |U| \leq |H(J)| \cdot |U|$. Since U is finite, we receive $|J| \leq |H(J)|$ which proves the claim. So we answered the problem positively. The generalization of Theorem 1.2 to arbitrary Hall families by P.Hall (Theorem 1.3) implies that we can generalize the assertion to arbitrary groups G, if only U is a finite subgroup of G. For a complete discussion of the algebraic problem see [vdW]. - Let us now prove the theorems of P.Hall and M.Hall.

Lemma 1.1 Let $F = (F(i) : i \in I)$ be a family, and let K be a finite subset of I such that $|K| = |F(K)|$. If F satisfies Hall's condition, then so does the family $(F \setminus K) \setminus F(K)$.

Proof. Let $J \subset\subset I \setminus K$. By assumption, we have $|F(K \cup J)| \geq |K \cup J|$. Hence $|F(J) \setminus F(K)| = |F(J \cup K)| - |F(K)| \geq |J \cup K| - |K| = |J \setminus K| = |J|$.

Theorem 1.2. A finite Hall family satisfies Hall's condition if and only if it has a marriage.

Proof. Let $H = (H(i) : i \in I)$ be a finite Hall family. We prove the nontrivial part of the theorem by induction on $|H|$. If $H = \emptyset$, then $\emptyset$ is a marriage of H. Now assume $|H| > 0$. Choose $i \in I$ and $a \in H(i)$. Clearly $|H(i)| \geq 1 = |\{i\}|$.

Case 1. There exists a $K \subseteq I \setminus \{i\}$ such that $a \in H(K)$ and $|K| = |H(K)|$.

Then $K \neq \emptyset$, $|(H \setminus K) \setminus H(K)| < |H|$, and the family $H \setminus K \setminus H(K)$ satisfies Hall's condition by Lemma 1.1. Conseqently it has a marriage f by the induction hypothesis. Further $|H \upharpoonright K| < |H|$ and $H \upharpoonright K \subseteq H$ satisfies Hall's condition, so it has a marriage g. Obviously $f \cup g$ is a marriage of H.

Case 2. $|K| < |H(K)|$ for each $K \subseteq I \setminus \{i\}$ with $a \in H(K)$.

Then $(H \setminus \{i\}) \setminus \{a\}$ satisfies Hall's condition and has a marriage f by the induction hypothesis. The function $g = f \cup \{(i,a)\}$ is a marriage of H.

Theorem 1.3. A Hall family satisfies Hall's condition if and only if it has a marriage.

Proof. Let $H = (H(i) : i \in I)$. To prove the nontrivial part of the theorem, call a matching f in H admissible if $H \setminus f$ $(= (H \setminus \mathrm{dom}\, f) \setminus \mathrm{rng}\, f)$ satisfies Hall's condition. By assumption, $\emptyset$ is admissible.

Claim 1. If $\mathcal{F}$ is a chain of admissible matchings, ordered by inclusion, then $f = \bigcup \mathcal{F}$ is admissible.

To prove the claim, let $J \subset\subset I \setminus \mathrm{dom}\, f$. $H(J)$ is finite and $\mathcal{F}$ is a chain, so there exists a $g \in \mathcal{F}$ such that $H(J) \setminus \mathrm{rng}\, g = H(J) \setminus \mathrm{rng}\, f$. Since $H \setminus g$ satisfies Hall's condition, we have $|H(J) \setminus \mathrm{rng}\, f| = |H(J) \setminus \mathrm{rng}\, g| \geq |J|$. This shows that f is admissible.

By Zorn's Lemma, there exists a maximal admissible matching f in H. If $G = H \setminus f$, then G satisfies Hall's condition. Obviously it suffices to prove

Claim 2. $G = \emptyset$.

Assume for contradiction that there is an $i \in \mathrm{dom}\, G$. Since $|G(i)| \geq |\{i\}| = 1$, we can choose an element a of $G(i)$.

Case 1. There exists a $K \subset\subset \mathrm{dom}\, G \setminus \{i\}$ with $a \in G(K)$ and $|G(K)| = |K|$.

Then $K \neq \emptyset$ and $G \upharpoonright K \subseteq G$ satisfies Hall's condition, which implies, by Theorem 1.2, that it has a marriage g. Now $|K| = |G(K)|$, so $g[K] = G(K)$ and $H \setminus (f \cup g) = G \setminus g = (G \setminus K) \setminus G(K)$, and Lemma 1.1 shows that $H \setminus (f \cup g)$ satisfies Hall's condition. We have shown that $f \cup g$ is admissible. But $f \subsetneq f \cup g$, which contradicts the maximality of f.

Case 2. $|K| < |G(K)|$ for each $K \subseteq I \setminus \{i\}$ with $a \in G(K)$.

Then $H \setminus (f \cup \{(i,a)\}) = (G \setminus \{i\}) \setminus \{a\}$ satisfies Hall's condition, and $f \subsetneq f \cup \{(i,a)\}$ yields again a contradiction to the choice of f. So $G = \emptyset$, and the theorem is proved.

Corollary 1.4. (Compactness Theorem)
If a Hall family has no marriage, then it has a finite subfamily which possesses no marriage.

We have called a subset K of dom F critical in F if $F \upharpoonright K \in \mathcal{M}$ and $f[K] = F(K)$ for every marriage f of $F \upharpoonright K$. Let H be a Hall family and $K \subset\subset \mathrm{dom}\, H$. Then, by Theorem 1.2, K is critical in H if and only if $H \upharpoonright K$ satisfies Hall's condition and $|H(K)| = |K|$.

Corollary 1.5. (Smallest counterexample)
If a Hall family $H = (H(i) : i \in I)$ has no marriage, then there exist a set $K \subset\subset I$ critical in H and an $i \in I \setminus K$ such that $H(i) \subseteq H(K)$.

Proof. If H has no marriage, then, by Theorem 1.3, $|H(L)| < |L|$ for some $L \subset\subset I$. Choose L in such a way that $|L|$ is minimal, and let $i \in L$ and $K = L \setminus \{i\}$. Then $H \upharpoonright K$ must fulfil Hall's condition, so $H \upharpoonright K \in \mathcal{M}$. Further $|H(K)| = |K|$. Consequently $H \upharpoonright K$ is critical and $H(i) \subseteq H(K)$.

Let H be a Hall family. The Theorem of Hall answers the question whether there exists a marriage of H. Now, if we assume that H has a marriage, another interesting question is: which women are always married? The answer is given by the following theorem.

Theorem 1.6.
Let $H = (H(i) : i \in I)$ be a Hall family. If H has a marriage, then the following conditions are equivalent for any finite set B:
(1) $B \subseteq \operatorname{rng} f$ for every marriage f of H.
(2) $B \subseteq H(K)$ for some $K \subset\subset I$ which is critical in H.

Proof. If f is a marriage of H and $B \subseteq H(K)$ for some critical K, then $\operatorname{rng} f \supseteq f[K] = H(K) \supseteq B$, so (2) implies (1). Now assume (1) and let $b \in B$. Then $H \setminus \{b\}$ has no marriage. By Hall's Theorem, there exists a $K_b \subset\subset I$ such that $|H(K_b) \setminus \{b\}| < |K_b|$. Since $H \in \mathcal{M}$, we have $|H(K_b)| = |K_b|$, and this implies $b \in H(K_b)$. Take $K = \bigcup\{K_b : b \in B\}$. Then $B \subseteq H(K)$ and clearly $H \upharpoonright K \in \mathcal{M}$. If f is an arbitrary marriage of $H \upharpoonright K$, then $f[K] = H(K)$, since $|H(K_b)| = |K_b|$ for each $b \in B$. So we have shown that K is critical in H.

A further interesting question for families which possess a marriage is the following: Are there marriages in which as many women as possible leave unmarried? To investigate this problem we call a set A a surplus with respect to the family F if $(F \setminus \emptyset) \setminus A$ has a marriage. A surplus $A \subseteq X$ is called a maximal surplus of X with respect to F if, for all $a \in X \setminus A$, $A \cup \{a\}$ is not a surplus with respect to F.

Let $F \in \mathcal{M}$ be a family and X be a set of women. Of course there exists a maximal surplus of X with respect to F if X is finite. But for infinite X this must not be the case. Take for example $F = \{(n, \omega) : n \in \omega\}$ and $X = \omega$. We can give a positive answer for Hall families.

Theorem 1.7. If H is a Hall family and if $B \subseteq X$ is a surplus of X with respect to H, then there exists a maximal surplus $A \supseteq B$ of X with respect to H.

Proof. Let $\mathcal{A} = \{A : B \subseteq A \subseteq X$ and A is a surplus with respect to $H\}$. Clearly $\mathcal{A} \neq \emptyset$, since $B \in \mathcal{A}$. To apply Zorn's Lemma, let $\mathcal{B} \subseteq \mathcal{A}$ be a $\subseteq$-chain and assume for contradiction that $\bigcup\mathcal{B} \notin \mathcal{A}$. Then $H \setminus \bigcup\mathcal{B}$ has no marriage. By Hall's Theorem, there exists a finite subset L of $\operatorname{dom} H$ such that $|F(L) \setminus \bigcup\mathcal{B}| < |L|$. Since $\mathcal{B}$ is a chain and $F(L)$ is finite, we have $F(L) \setminus \bigcup\mathcal{B} = F(L) \setminus A$ for some $A \in \mathcal{B}$. This implies that A is no surplus with respect to F, which contradicts $A \in \mathcal{B}$. Zorn's Lemma yields a $\subseteq$-maximal $A \in \mathcal{A}$, which is a maximal surplus of X with respect to H.

Theorem 1.8. If A is a finite maximal surplus of X with respect to a Hall family H and $B \subseteq X$ is a surplus with respect to H, then $|B| \leq |A|$.

Proof. It suffices to prove Theorem 1.8 for finite sets B. By assumption, $H \setminus A$ has a marriage and $H \setminus (A \cup \{a\})$ has no marriage for every $a \in B \setminus A$. Hence, by Theorem 1.6, there is a $K \subset\subset \operatorname{dom} H$ with $|H(K) \setminus A| = |K|$ and $B \setminus A \subseteq H(K)$. $H \setminus B$ has a marriage, so $|H(K) \setminus B| \geq |K|$. Therefore

$$\begin{aligned} |K| + |A| &= |H(K) \setminus A| + |A| = |H(K) \cup A| \\ &\geq |H(K) \cup B| = |H(K) \setminus B| + |B| \\ &\geq |K| + |B| \end{aligned}$$

Since K is finite, $|B| \leq |A|$, and the theorem is proved.

Finally we investigate the following problem: If the family $F = (F(i) : i \in I)$ has no marriage, does there exist a matching in F which matches as many men as possible?

Remember that we have called a set $J \subseteq I$ matchable in F if $F \restriction J \in \mathcal{M}$. Of course we say in this case that $F \restriction J$ is a matchable subfamily. A matchable set $J \subseteq I$ is called maximal matchable in F if $F \restriction J \cup \{i\}$ has no marriage for each $i \in I \setminus J$. We call the cardinal

$$d(F) = \min\{|I \setminus J| : J \text{ is matchable in } F\}$$

the defect of F.

Not every family possesses a maximal matchable subfamily: consider the family $\{(\alpha,\omega) : \alpha < \omega_1\}$. But for Hall families we can give a positive answer again.

Theorem 1.9. If $H = (H(i) : i \in I)$ is a Hall family and $J \subseteq I$ is matchable in H, then there exists a subset M of I such that $J \subseteq M$ and M is maximal matchable in H.

Proof. Let $\mathcal{A} = \{L : J \subseteq L \subseteq I$ and L is matchable in $H\}$. Clearly $\mathcal{A} \neq \emptyset$, and if $\mathcal{B} \subseteq \mathcal{A}$ is a $\subseteq$-chain in $\mathcal{A}$, then Hall's Theorem implies that $\cup\mathcal{B} \subset \mathcal{A}$. By Zorn's Lemma, there exists a $\subseteq$-maximal element M of $\mathcal{A}$ which has the desired properties.

§2. Frames

In this section we want to investigate a class of families which are "almost" Hall families. To get an idea of what we are going to do consider the family F_0, defined by

$\operatorname{dom} F_0 = \omega + 1$,

$F_0(n) = n+2$ for each $n \in \omega$, $F_0(\omega) = \omega$.

Clearly F_0 is not a Hall family. But we can associate with F_0 a Hall family H_0 as follows:

$\operatorname{dom} H_0 = \operatorname{dom} F_0$,

$H_0(n) = n+2$ for each $n \in \omega$, $H_0(\omega) = \{2,9\}$.

Further we define a function $r_0 : \operatorname{dom} H_0 \to On$ by $r_0(n) = 0$ for each $n \in \omega$ and $r_0(\omega) = 1$, a so-called rank function for H. Consider the man ω of rank 1. If f is a matching of all men of rank 0, then $|H_0(\omega) \setminus \operatorname{rng} f| \leq 1$, and further there exists a matching g of all men of rank 0 with $|H_0(\omega) \setminus \operatorname{rng} g| = 1$.

Hence $H_0(\omega)$ contains a maximal subset of $F_0(\omega)$ of women which remain unmarried after we have married $\{i \in \operatorname{dom} H_0 : r_0(i) < 1\}$. Therefore we can restrict our attention to $H_0(i)$ if we want to match in F additionally the man i with $r_0(i) = 1$.

We want to characterize this situation in general. First we need the following lemma.

Lemma 2.1.
Let $H = (H(i) : i \in I)$ be a Hall family and B a finite subset of the set X. If H has a marriage, then the following conditions are equivalent:

(1) There is a maximal surplus $A \subseteq B$ of X with respect to H.

(2) For each $a \in X$ and each $L \subseteq I$, there exists a $K \subset\subset I$ such that $a \in H(K \cup L) \cup B$ and $|H(K) \setminus (H(L) \cup B)| \leq |K \setminus L|$.

Proof. Assume (1) and let $a \in X$ and $L \subseteq I$. If $a \in A$, then take $K = \emptyset$. So assume that $a \in X \setminus A$. Since $a \in \operatorname{rng} f$ for every marriage f of $H \setminus A$, there exists, by Theorem 1.6, a finite set $K \subseteq I$ which is critical in $H \setminus A$ such that $a \in H(K) \cup A$. Then $|H(K) \setminus A| = |K|$ and $|H(K \cap L) \setminus A| \geq |K \cap L|$. Therefore $|H(K) \setminus (H(L) \cup B)| \leq |H(K) \setminus (H(K \cap L) \cup A)|$ $= |H(K) \setminus A| - |H(K \cap L) \setminus A| \leq |K| - |K \cap L| = |K \setminus L|$. So we get (2).

To prove the converse, assume (2) and let A be a maximal surplus of B with respect to H. We want to show that A is a maximal surplus of X with respect to H. It is sufficient to find for each $a \in X \setminus A$ an $M \subset\subset I$ such that $a \in H(M)$ and $|H(M) \setminus A| \leq |M|$. By assumption on A and by Theorem 1.6, there exists an $L \subset\subset I$ such that $B \setminus A \subseteq H(L) \setminus A$ and $|H(L) \setminus A| = |L|$. Now (2) yields a $K \subset\subset I$ with $a \in H(K \cup L) \cup B = H(K \cup L) \cup A$ and $|H(K) \setminus (H(L) \cup B)| \leq |K \setminus L|$. Take $M = K \cup L$. Then $|H(M) \setminus A| =$ $= |H(K) \setminus (H(L) \cup A)| + |H(L) \setminus A| \leq |K \setminus L| + |L| = |M|$, and the lemma is proved.

These considerations lead to the following definition.

Definition. Let F and H be families with $\operatorname{dom} F = \operatorname{dom} H$ and $H(i) \subset\subset F(i)$ for every $i \in \operatorname{dom} H$, and let $r : \operatorname{dom} H \to On$ be a function.

H is called a frame of F if for each $i \in \operatorname{dom} H$, each $L \subseteq \operatorname{dom} H$ with $i \in L$, and each $a \in F(i)$ there is a $K \subset\subset \{j : r(j) < r(i)\}$ such that $a \in H(K \cup L)$ and $|H(K) \setminus H(L)| \leq |K \setminus L|$. r is said to be a rank function for H. We say that F has a frame if there is a frame of F.

Clearly in our example H_0 is a frame of F_0. - Let H be a frame of F and r a rank function for H. Let $r(i) = \alpha$, $I_\alpha = \{j : r(j) < \alpha\}$ and suppose that $H \upharpoonright I_\alpha$ has a marriage. Then, by Lemma 2.1 and the definition of a frame, there exists a maximal surplus $A \subseteq H(i)$ of $F(i)$ with respect to $H \upharpoonright I_\alpha$. Hence we get by Theorem 1.8:

(i) If $B \subseteq F(i)$ is a surplus with respect to $H \upharpoonright I_\alpha$, then there is a surplus $A \subseteq H(i)$ with respect to $H \upharpoonright I_\alpha$ such that $|B| \leq |A|$.

Since this holds for every α and i, we can hope that:

(ii) If $B \subseteq F(i)$ is a surplus with respect to $F \upharpoonright I_\alpha$, then there is a surplus $A \subseteq H(i)$ with respect to $H \upharpoonright I_\alpha$ such that $|B| \leq |A|$.

This describes the situation in our example. We are able to prove (ii) after the proof of Theorem 2.3.

We can construct families which have a frame as follows. Take a Hall family H and a function $r : \operatorname{dom} H \to On$. Then extend $H(i)$ to $F(i)$ by joining only those women a which have the following property:
If $i \in L \subseteq \operatorname{dom} H$, then there is a $K \subset\subset \{j : r(j) < r(i)\}$ such that $a \in H(L \cup K)$ and $|H(K) \setminus H(L)| \leq |K \setminus L|$.

Roughly speaking, we admit for $F(i)$ only women who are already promised to a man of lower rank. Hence families which have a frame are nearly Hall families. Especially we will show that every critical family can be constructed in this way from a critical Hall family. In Section 7 we prove that they are just the stable families which can be constructed in this way.

If H is a family with rank function r and $L \subset\subset \operatorname{dom} H$, then define $\underline{r(L)} := \max\{r(i) + 1 : i \in L\}$.

Lemma 2.2. Let H be a frame of F and r be a rank function for H. If $\emptyset \neq J \subset\subset \operatorname{dom} H$, $D \subset\subset F(J)$, and $J \subseteq L \subseteq \operatorname{dom} H$, then there is a $K \subset\subset \operatorname{dom} H$ such that $D \subseteq H(K \cup L)$, $|H(K) \setminus H(L)| \leq |K \setminus L|$, and $r(K) < r(J)$.

Proof. We use induction on $|D|$. If $|D| = 0$, then let $K = \emptyset$. So assume $|D| > 0$ and let $d \in D$. By the induction hypothesis, there is a $K_1 \subset\subset \operatorname{dom} H$ with $D \setminus \{d\} \subseteq H(K_1 \cup L)$, $|H(K_1) \setminus H(L)| \leq |K_1 \setminus L|$, and $r(K_1) < r(J)$. Let $i \in J \subseteq K_1 \cup L$ with $d \in F(i)$. By the definition of a frame, there is a set $K_2 \subset\subset \{j : r(j) < r(i)\}$ such that $d \in H(K_2 \cup K_1 \cup L)$ and $|H(K_2) \setminus H(K_1 \cup L)| \leq |K_2 \setminus (K_1 \cup L)|$. Put $K = K_1 \cup K_2$. Then $D \subseteq H(K \cup L)$, $r(K) < r(J)$, and $|H(K) \setminus H(L)| =$
$= |H(K_2) \setminus H(K_1 \cup L)| + |H(K_1) \setminus H(L)| \leq |K_2 \setminus (K_1 \cup L)| + |K_1 \setminus L| = |K \setminus L|$.

Theorem 2.3. Let H be a frame of F. If A is a surplus with respect to F, then A is a surplus with respect to H.

Proof. Let r be a rank function for H, and let f be a marriage of $F \setminus A$. Suppose that $H \setminus A$ has no marriage. Then, by Hall's Theorem, there is a set $L_0 \subset\subset \operatorname{dom} H$ such that

$|H(L_0) \setminus A| < |L_0|$.

By recursion, we will define finite sets L_n such that for $n > 0$

(1) $|H(L_n) \setminus A| < |L_n|$,

(2) $f[L_{n-1}] \subseteq H(L_n)$,

(3) $r(L_{n+1} \setminus L_n) < r(L_n \setminus L_{n-1})$.

Then (3) is a contradiction to the fact that On is well-founded. Suppose that L_n is already chosen for $n \geq 0$. Let $J = \{i \in L_n : f(i) \notin H(L_n)\}$. Since $|H(L_n) \setminus A| < |L_n|$, we have $J \neq \emptyset$. Hence, by Lemma 2.2, there is a set $K \subset\subset \operatorname{dom} H$ with $f[J] \subseteq H(K \cup L_n)$, $|H(K) \setminus H(L_n)| \leq |K \setminus L_n|$ and $r(K) < r(J)$. Let $L_{n+1} = K \cup L_n$. Then

1. $|H(L_{n+1}) \setminus A| = |H(K) \setminus (H(L_n) \cup A)| + |H(L_n) \setminus A| < |K \setminus L_n| + |L_n| = |L_{n+1}|$.
2. $f[L_n] \subseteq H(L_{n+1})$.
3. Let $n > 0$. Since $f[L_{n-1}] \subseteq H(L_n)$, we have $J \subseteq L_n \setminus L_{n-1}$. Therefore $r(L_{n+1} \setminus L_n) \leq r(K) < r(J) \leq r(L_n \setminus L_{n-1})$.

Now we can prove property (ii) of page 106. Let $B \subseteq F(i)$ be a surplus with respect to $F \upharpoonright I_\alpha$. $H \upharpoonright I_\alpha$ is a frame of $F \upharpoonright I_\alpha$. Hence, by Theorem 2.3, B is a surplus with respect to $H \upharpoonright I_\alpha$. Then, by (i), there is a surplus $A \subseteq H(i)$ with respect to $H \upharpoonright I_\alpha$ such that $|B| \leq |A|$.

We can use frames to give a necessary and sufficient condition for the existence of a marriage for a countable family.

<u>Definition</u>. We call H a <u>frame in F</u> if H is a frame of $F \upharpoonright J$ for some $J \subseteq \operatorname{dom} F$.

If F has a marriage, then, by Theorem 2.3, the following condition is satisfied for F:

(f) Every frame in F fulfils Hall's condition.

To show that every countable F with property (f) has a marriage we need some facts about frames in F.

<u>Lemma 2.4</u>. If $\mathcal{G}$ is a chain of frames in F, ordered by inclusion, then $H = \bigcup \mathcal{G}$ is a frame in F.

<u>Proof</u>. For each $G \in \mathcal{G}$ let s_G be a rank function for G. We define the function r by

$$r(i) = \min\{s_G(i) : i \in \operatorname{dom} G \text{ and } G \in \mathcal{G}\}.$$

We want to show that r is a suitable rank function for H. Let $i \in \operatorname{dom} H$, $i \in L \subseteq \operatorname{dom} H$, and $a \in F(i)$. Take $G \in \mathcal{G}$ with $s_G(i) = r(i)$ and put $L' = L \cap \operatorname{dom} G$. Then there is some $K \subset\subset \{j : s_G(j) < s_G(i)\} \subseteq \{j : r(j) < r(i)\}$ such that $a \in G(K \cup L') \subseteq H(K \cup L)$ and $|H(K) \setminus H(L)| \leq |G(K) \setminus G(L')| \leq |K \setminus L'| = |K \setminus L|$.

Since $\emptyset$ is a frame in F, we get by Zorn's Lemma:

Corollary 2.5. There is a maximal frame in F.

Lemma 2.6. Let H be a frame in F and $i_0 \notin \operatorname{dom} H$. If there exists a finite maximal surplus C of $F(i_0)$ with respect to $H \upharpoonright J$ for some $J \subseteq \operatorname{dom} H$, then there is a frame G such that $H \subseteq G$ and $i_0 \in \operatorname{dom} G$.

Proof. Let $G = H \cup \{(i_0, C)\}$. We want to show that G is a frame in F.

Let $r : \operatorname{dom} H \to \alpha$ be a rank function for H. Put $s = r \cup \{(i_0, \alpha)\}$ and let $i \in \operatorname{dom} G$ and $L \subseteq \operatorname{dom} G$ such that $i \in L$ and $a \in F(i)$. We must find a set $K \subset\subset \{j : s(j) < s(i)\}$ such that $a \in G(K \cup L)$ and $|G(K) \setminus G(L)| \leq |K \setminus L|$. Let $L' = L \setminus \{i_0\}$. If $s(i) < \alpha$, then there is a set $K \subset\subset \{j : r(j) < r(i)\} = \{j : s(j) < s(i)\}$ such that $a \in H(K \cup L') \subseteq G(K \cup L)$ and $|G(K) \setminus G(L)| \leq |H(K) \setminus H(L')| \leq |K \setminus L'| = |K \setminus L|$, and we are done. So assume that $s(i) = \alpha$, i.e. $i = i_0$. $C \subseteq G(i_0)$ is a maximal surplus of $F(i_0)$ with respect to $H \upharpoonright J$. Therefore, by Lemma 2.1, there is a set $K \subset\subset J \subseteq \{j : s(j) < \alpha\}$ such that $a \in H(K \cup L') \cup G(i_0) = G(K \cup L)$ and $|G(K) \setminus G(L)| = |H(K) \setminus (H(L') \cup G(i_0))| \leq |K \setminus L'| = |K \setminus L|$.

Corollary 2.7. Let H be a maximal frame in F and $i \notin \operatorname{dom} H$. If $J \subseteq \operatorname{dom} H$ and if B is a finite surplus with respect to $H \upharpoonright J$, then there is a $b \in F(i) \setminus B$ such that $B \cup \{b\}$ is a surplus with respect to $H \upharpoonright J$.

Proof. Suppose there is no such b. Then B is a maximal surplus of $F(i) \cup B$ with respect to $H \upharpoonright J$. Let $C \subset\subset F(i)$ be a surplus of $F(i) \cup B$ with respect to $H \upharpoonright J$. By Theorem 1.8, we have $|C| \leq |B|$. Hence there is a finite maximal surplus C of $F(i)$ with respect to $H \upharpoonright J$. By Lemma 2.6, there is a frame $G \supseteq H$ with $i \in \operatorname{dom} G$. This contradicts the maximality of H.

The main theorem of this chapter is the following

Theorem 2.8. Let H be a maximal frame in F, let A be finite, and let J be countable such that $F \setminus J$ has a marriage. If A is a surplus with respect to H, then A is a surplus with respect to F.

Proof. We call a finite matching g in F admissible if $(H \setminus A) \setminus g$

has a marriage. $H\setminus A$ has a marriage, hence $\emptyset$ is admissible.

Claim. If g is admissible and $K\subset\subset \operatorname{dom} F$, then there is an admissible matching $h\supseteq g$ with $K\subseteq \operatorname{dom} h$.

We prove the claim by induction on $|K\setminus \operatorname{dom} g|$. If $K\subseteq \operatorname{dom} g$, then let $h = g$. Otherwise let $i\in K\setminus \operatorname{dom} g$. Then, by the induction hypothesis, there is an admissible $h_1\supseteq g$ with $K\setminus\{i\}\subseteq \operatorname{dom} h_1$. If $i\in \operatorname{dom} h_1$, then let $h = h_1$. Otherwise we distinguish two cases.

Case 1. $i\in \operatorname{dom} H$.
Let f be a marriage of $(H\setminus A)\setminus h_1$ and let $h = h_1\cup\{(i,f(i))\}$.

Case 2. $i\notin \operatorname{dom} H$.
$A\cup \operatorname{rng} h_1$ is a surplus with respect to $H\setminus \operatorname{dom} h_1$.
Hence, by Corollary 2.7, there is an $a\in F(i)\setminus(A\cup \operatorname{rng} h_1)$ such that $A\cup \operatorname{rng} h_1\cup\{a\}$ is a surplus with respect to $H\setminus \operatorname{dom} h_1$. Then $h = h_1\cup\{(i,a)\}$ has the desired properties, and the claim is proved.

Now let f be a marriage of $F\setminus J$ and $J_n\subset\subset J$ for each $n\in\omega$ such that $\bigcup\{J_n : n\in\omega\} = J$. We choose admissible matchings g_n, $n\in\omega$, as follows:
Let $g_0 = \emptyset$ and suppose that g_n is already chosen. Take $K_n = J_n\cup\{i : f(i)\in \operatorname{rng} g_n\cup A\}$, and let $g_{n+1}\supseteq g_n$ with $K_n\subseteq \operatorname{dom} g_{n+1}$.

If $g = \bigcup\{g_n : n\in\omega\}$, then $f\restriction(\operatorname{dom} f\setminus \operatorname{dom} g)\cup g$ is a marriage of $F\setminus A$.

Theorem 2.9.
If F is a countable family, then the following propositions are equivalent:
(1) F has a marriage.
(2) Every frame in F satisfies Hall's condition.

Proof. (1) implies (2) by Theorem 2.3. For the proof of the converse let H be a maximal frame in F. By (2) and Hall's Theorem, H has a marriage. If $J = \operatorname{dom} F$, then $F\setminus J = \emptyset$ has a marriage. Hence, by Theorem 2.8, F has a marriage.

Theorem 2.10.
The following propositions are equivalent:
(1) Every frame in F has a marriage.
(2) If $F\restriction J\in\mathcal{M}$ and $i\in \operatorname{dom} F$, then $F\restriction J\cup\{i\}\in\mathcal{M}$.

Proof. Assume (1), let $J \subseteq \text{dom}\, F$ such that $F \upharpoonright J \in \mathcal{M}$, and let $i \in \text{dom}\, F$. Take a maximal frame H in $F \upharpoonright J \cup \{i\}$. By assumption, H has a marriage. Therefore, by Theorem 2.8, $F \upharpoonright J \cup \{i\}$ has a marriage.

To prove the converse, let H be a frame in F and $r : \text{dom}\, H \to \gamma$ a rank function for H. Put $I_\alpha = \{i : r(i) < \alpha\}$. Since $H \upharpoonright I_\gamma = H$, it is sufficient to show that $H \upharpoonright I_\alpha$ has a marriage for every $\alpha \leq \gamma$. We prove this by induction on α. Assume that $H \upharpoonright I_\beta \in \mathcal{M}$ for every $\beta < \alpha$. Let $\beta < \alpha$ and $M \subset\subset I_{\beta+1}$. By the Compactness Theorem 1.4, it is sufficient to show that $H \upharpoonright I_\beta \cup M$ has a marriage. We prove this by induction on $|M \setminus I_\beta|$. If $M \subseteq I_\beta$, this holds by assumption. Otherwise let $i \in M \setminus I_\beta$. By the induction hypothesis, $H \upharpoonright I_\beta \cup (M \setminus \{i\}) \in \mathcal{M}$. Therefore $F \upharpoonright I_\beta \cup (M \setminus \{i\}) \in \mathcal{M}$ and consequently, by (2), $F \upharpoonright I_\beta \cup M$ has a marriage. $H \upharpoonright I_\beta \cup M$ is a frame of $F \upharpoonright I_\beta \cup M$, hence $H \upharpoonright I_\beta \cup M$ has a marriage by Theorem 2.3.

Now let us turn our attention to critical families.

Theorem 2.11. If $F \setminus A$ is critical for some finite set A, then F has a frame.

Proof. Let H be a maximal frame in F. Since $F \setminus A$ has a marriage, $H \setminus A$ has a marriage by Theorem 2.3. Suppose $\text{dom}\, H \neq \text{dom}\, F$ and let $i \in \text{dom}\, F \setminus \text{dom}\, H$. By Corollary 2.7, there is an $a \in F(i) \setminus A$ such that $H \setminus (A \cup \{a\})$ has a marriage. Then, by Theorem 2.8, there exists a marriage f of $F \setminus A$ with $a \notin \text{rng}\, f$, which contradicts the assumption that $F \setminus A$ is critical.

From this theorem we get the following characterization of critical sets:

Corollary 2.12. A family is critical if and only if it has a critical frame.

Proof. If F is critical, then, by Theorem 2.11, F has a frame H. Since $F \in \mathcal{M}$, $H \in \mathcal{M}$ by Theorem 2.3. Further every marriage of H is a marriage of F, hence H must be critical.

Conversely let H be a critical frame of F. Clearly $F \in \mathcal{M}$, since $H \in \mathcal{M}$. Further $F(\text{dom}\, F) = H(\text{dom}\, H)$ by the definition of a frame. Suppose that $F \setminus \{a\}$ has a marriage for some $a \in F(\text{dom}\, F)$. Then, by Theorem 2.3,

$H\setminus\{a\}$ has a marriage. This is impossible, since H is critical. So we have shown that F is critical.

We can use frames to reprove some important facts about critical families. In this chapter we will use the following special case of Theorem II.2.10.

Theorem 2.13. Let A be finite, and let K be critical in $F\setminus A$. If $B\subseteq F(K)\cup A$ such that $(F\upharpoonright K)\setminus B$ has a marriage, then $|B|\leq|A|$.

Proof. By Theorem 2.11, there exists a frame H of $F\upharpoonright K$. $F\upharpoonright K\setminus A\in\mathcal{M}$ and so, by Theorem 2.3, $H\setminus A\in\mathcal{M}$. Since $F\upharpoonright K\setminus A$ is critical, A must be a maximal surplus of $F(K)\cup A$ with respect to H. Further, since $F\upharpoonright K\setminus B\in\mathcal{M}$, $H\setminus B$ has a marriage by Theorem 2.3 again. Therefore B is a surplus of $F(K)\cup A$ with respect to H. So we get by Theorem 1.8 that $|B|\leq|A|$.

We can give a new proof of the Main Lemma (Theorem 2.11 of Chapter II).

Theorem 2.14. If F has a marriage, then there is a set M critical in F such that $F\setminus\{a\}$ has a marriage for each $a\notin F(M)$.

Proof. Let H be a maximal frame in F. By Theorem 2.3, H has a marriage. Let

$$M = \cup\{L\subset\subset \operatorname{dom} H : |H(L)|\leq|L|\}.$$

Claim 1. $F(M) = H(M)$.

Let $i\in M$, and let $a\in F(i)$. By the definition of M, there is an $L\subset\subset\operatorname{dom} H$ such that $i\in L$ and $|H(L)|\leq|L|$. Now, by the definition of a frame, there is a set $K\subset\subset\operatorname{dom} H$ such that $a\in H(K\cup L)$ and $|H(K)\setminus H(L)|\leq|K\setminus L|$. So $|H(K\cup L)|\leq|K\cup L|$, which implies $K\cup L\subseteq M$ and $a\in H(K\cup L)\subseteq H(M)$.

Claim 2. $F\upharpoonright M$ is critical.

Suppose that there is a $b\in F(M) = H(M)$ and a marriage g of $F\upharpoonright M\setminus\{b\}$. Let h be a marriage of H. Then $h[M] = H(M) = F(M)$ by the definition of M, and $g\cup(h\upharpoonright(\operatorname{dom} H\setminus M))$ is a marriage of $(F\upharpoonright\operatorname{dom} H)\setminus\{b\}$. Consequently, by Theorem 2.3, $H\setminus\{b\}$ has a marriage

f. Since $f[M] \neq H(M)$, this is a contradiction to the definition of M.

Now let $a \notin F(M)$.

Claim 3. $F \setminus \{a\}$ has a marriage.

Since $H \in \mathcal{M}$ and $a \notin F(M) = H(M)$, $H \setminus \{a\}$ has a marriage by Theorem 1.6. Theorem 2.8 implies that $F \setminus \{a\}$ has a marriage.

§3. Critical families

The main result of this chapter is the one that, for every family F, the following conditions are equivalent.

(b) If $F \upharpoonright J$ has a marriage, then $F \upharpoonright J \cup \{i\}$ has a marriage for every $i \in \operatorname{dom} F$.

(c) For any K critical in F and any $i \notin K$, $F(i) \nsubseteq F(K)$.

Clearly, if (b) holds, then (c) holds.

To prove the other direction we need some facts about critical sets which we already proved in Section II.2. The following Facts 1, 2 and 3 are immediate consequences of the definition of a critical set.

Fact 1. If $\mathcal{K}$ is a chain of critical sets in F, then $\bigcup \mathcal{K}$ is critical in F.

Proof. Lemma II.2.3.

$\emptyset$ is critical in F. Hence by Zorn's Lemma we get:

Corollary 3.1.

(i) For each K critical in F, there is a set $L \supseteq K$ which is maximal critical in F.

(ii) There is a maximal critical set L in F.

Fact 2. If K is critical in F and L is critical in $F \setminus F(K)$, then $K \cup L$ is critical in F.

Proof. Lemma II.2.6.

Fact 3. If K is critical in F and X is a set, then there is a set $L \subseteq K$ critical in $F \setminus X$ such that $F(K) \cup X = F(L) \cup X$.

Proof. Lemma II.2.5.

Corollary 3.2. If K is maximal critical in F and L is critical in F, then $F(L) \subseteq F(K)$.

Proof. By Fact 3, there is an $N \subseteq L$ critical in $F \setminus F(K)$ such that $F(N) \cup F(K) = F(L) \cup F(K)$. $N \cup K$ is critical in F by Fact 2, and K is maximal critical, so $N = \emptyset$ and $F(L) \subseteq F(K)$.

More work has to be done to prove Fact 4.

Fact 4. Let A be finite, and let K be critical in $F \setminus A$. If $B \subseteq F(K) \cup A$ such that $(F \upharpoonright K) \setminus B$ has a marriage, then $|B| \leq |A|$.

Proof. Theorem II.2.10 or Theorem 2.13.

Corollary 3.3. Let F be a family, let $i \in \operatorname{dom} F$ and $a \in F(i)$. If $F(i) \subseteq F(\operatorname{dom} F \setminus \{i\}) \cup \{a\}$ and $F \setminus \{i\} \setminus \{a\}$ is critical, then F is critical.

Proof. Let f be a marriage of $F \setminus \{i\} \setminus \{a\}$. Then $g = f \cup \{(i,a)\}$ is a marriage of F. Now let g be an arbitrary marriage of F. Put $K = \operatorname{dom} F \setminus \{i\}$, $A = \{a\}$, and $B = F(\operatorname{dom} F) \setminus g[K] \supseteq \{g(i)\}$. Then $B \subseteq F(K) \cup A$ and $F \upharpoonright K \setminus B$ has a marriage, which implies, by Fact 4, that $|B| \leq |\{a\}|$. Hence $B = \{g(i)\}$ and $\operatorname{rng} g = F(\operatorname{dom} F)$, and we have shown that F is critical.

Facts 1 - 4 are the most important facts about critical sets. We want to show that we can use these facts to give a third proof of the Main Lemma (II.2.11 and Theorem 2.14).

Theorem 3.4. (Main Lemma)
If F has a marriage, then there is a set K critical in F such that $F \setminus \{a\}$ has a marriage for each $a \notin F(K)$.

Proof. Let K be maximal critical in F, and let f be a marriage of F. Assume that $F \setminus \{a\}$ has no marriage for some $a \notin F(K)$. We will choose pairwise different i_n, $n \in \omega$, such that

$f(i_0) = a$ and $f(i_{n+1}) \in F(i_n)$ for $n \in \omega$.
Then $g = (f \setminus \{(i_n, f(i_n)) : n \in \omega\}) \cup \{(i_n, f(i_{n+1})) : n \in \omega\}$ is a marriage of F with $a \notin \operatorname{rng} g$. This is a contradiction to the assumption that $F \setminus \{a\} \notin \mathcal{M}$.

Since $F \setminus \{a\}$ has no marriage, there is an $i_0 \in \operatorname{dom} F$ with $f(i_0) = a$. Let $K_0 = K$ and suppose that we have already chosen $\{i_k : k \leq n\}$ and K_n maximal critical in $F_n = F \setminus f \restriction \{i_k : k < n\}$ such that $f(i_n) \notin F(K_n)$. Since $f(i_n) \notin F(K_n)$, K_n is critical in F_{n+1}. By Corollary 3.1, there is a $K_{n+1} \supseteq K_n$ maximal critical in F_{n+1}.

Claim 1. $F_n(i_n) \not\subseteq F_n(K_{n+1}) \cup \{f(i_n)\}$.

Suppose that $F_n(i_n) \subseteq F_n(K_{n+1}) \cup \{f(i_n)\}$. Then, by Corollary 3.3, $K_{n+1} \cup \{i_n\}$ is critical in F_n. Since $K_n \subseteq K_{n+1}$ and $i_n \notin K_n$, this is a contradiction to the maximality of K_n.

Let $b \in F_n(i_n) \setminus (F_n(K_{n+1}) \cup \{f(i_n)\})$ and note that $b \notin F(K_{n+1})$.

Claim 2. There is an i_{n+1} with $f(i_{n+1}) = b$.

Suppose that $b \notin \operatorname{rng} f$. Then
$$g = (f \setminus \{(i_k, f(i_k)) : k \leq n\}) \cup \{(i_k, f(i_{k+1})) : k < n\} \cup \{(i_n, b)\}$$
is a marriage of F with $a \notin \operatorname{rng} g$. This is a contradiction since, by assumption, $F \setminus \{a\}$ has no marriage.

Corollary 3.5.
The following propositions are equivalent:
(b) If $F \restriction J$ has a marriage and $i \in \operatorname{dom} F$, then $F \restriction J \cup \{i\}$ has a marriage.
(c) If K is critical in F and $F(i) \subseteq F(K)$, then $i \in K$.

Proof. Assume that K is critical in F and $F(i) \subseteq F(K)$. By (b), $F \restriction K \cup \{i\}$ has a marriage f. Since $F(i) \subseteq F(K) = f[K]$, we have $i \in K$.

For the proof of the converse let $F \restriction J \in \mathcal{M}$ and $i \in \operatorname{dom} F$ and assume that $i \notin J$. By the Main Lemma, there is a set $K \subset J$ critical in F such that $F \restriction J \setminus \{a\}$ has a marriage for each $a \notin F(K)$. Since $i \notin K$, and so $F(i) \not\subseteq F(K)$ by (c), we can choose an $a \in F(i) \setminus F(K)$ and a marriage f of $F \restriction J \setminus \{a\}$. Then $f \cup \{(i,a)\}$ is a marriage of $F \restriction J \cup \{i\}$.

If F is a countable family, then from Theorem 2.9 and Theorem 2.10 we get that F has a marriage if and only if (b) holds. We want to give a direct proof for this fact and use the following abbreviation.

Definition. A family F is called c-good if, for each $J \subseteq \operatorname{dom} F$ and each $i \in \operatorname{dom} F$, $F \restriction J \in \mathcal{M}$ implies $F \restriction J \cup \{i\} \in \mathcal{M}$.

Lemma 3.6. If F is c-good and K is maximal critical in F, then $F \setminus \{a\}$ is c-good for every $a \notin F(K)$.

Proof. Suppose that $F \restriction J \setminus \{a\}$ has a marriage, $a \notin F(K)$, and let $i \in \operatorname{dom} F$. We have to show that $(F \restriction J \cup \{i\}) \setminus \{a\}$ has a marriage. Since F is c-good, $F \restriction J \cup \{i\} \in \mathcal{M}$. By the Main Lemma, there is an L critical in $F \restriction J \cup \{i\}$ such that $(F \restriction J \cup \{i\}) \setminus \{b\}$ has a marriage for every $b \notin F(L)$. We have $F(L) \subseteq F(K)$ by Corollary 3.2, and therefore $(F \restriction J \cup \{i\}) \setminus \{a\}$ has a marriage.

Corollary 3.7. If F is c-good and $i \in \operatorname{dom} F$, then there is an $a \in F(i)$ such that $F \setminus \{i\} \setminus \{a\}$ is c-good.

Proof. Let K be maximal critical in $F \setminus \{i\}$. Since F is c-good, $F(i) \not\subseteq F(K)$. Let $a \in F(i) \setminus F(K)$. By Lemma 3.6, $F \setminus \{i\} \setminus \{a\}$ is c-good.

Theorem 3.8. Let A be finite, and let J be countable. If $F \setminus A$ is c-good and $F \setminus J$ has a marriage, then $F \setminus A$ has a marriage.

Proof. We follow the proof of Theorem 2.8 and call a finite matching g in F admissible if $(F \setminus A) \setminus g$ is c-good. First observe that $\emptyset$ is admissible, since $F \setminus A$ is c-good.

Claim. If g is admissible and $K \subset\subset \operatorname{dom} F$, then there is an admissible $h \supseteq g$ with $K \subseteq \operatorname{dom} h$.

We prove the claim by induction on $|K \setminus \operatorname{dom} g|$. If $|K \setminus \operatorname{dom} g| = 0$, then let $h = g$. Otherwise let $i \in K \setminus \operatorname{dom} g$. By the induction hypothesis, there is an admissible $h_1 \supseteq g$ with $K \setminus \{i\} \subseteq \operatorname{dom} h_1$. If $i \in \operatorname{dom} h_1$, then let $h = h_1$. Otherwise, since $(F \setminus A) \setminus h_1$ is c-good, there is an $a \in F(i) \setminus (A \cup \operatorname{rng} h_1)$ such that, by Corollary 3.7, $(F \setminus A) \setminus (h_1 \cup \{(i,a)\})$ is c-good. $h = h_1 \cup \{(i,a)\}$ has the desired properties.

To finish the proof let f be a marriage of $F \setminus J$, and let $J_n \subset\subset J$, $n \in \omega$, with $\bigcup\{J_n : n \in \omega\} = J$. We choose a chain $(g_n : n \in \omega)$ of admissible matchings as follows:

Let $g_0 = \emptyset$ and suppose that g_n is already chosen. Take $K_n = J_n \cup \{i : f(i) \in \operatorname{rng} g_n \cup A\}$ and let $g_{n+1} \supseteq g_n$ be admissible with $K_n \subseteq \operatorname{dom} g_{n+1}$. If $g = \bigcup\{g_n : n \in \omega\}$, then $f \upharpoonright (\operatorname{dom} f \setminus \operatorname{dom} g) \cup g$ is a marriage of $F \setminus A$.

Corollary 3.9.
If F is a countable family, then the following propositions are equivalent:
(a) F has a marriage.
(b) If $F \upharpoonright J$ has a marriage, then $F \upharpoonright J \cup \{i\}$ has a marriage for every $i \in \operatorname{dom} F$.

We can use these results to give a description of a maximal surplus of a set.

Theorem 3.10. If A is a maximal surplus of X with respect to F, then there is a set K critical in $F \setminus A$ such that $X \subseteq F(K) \cup A$.

Proof. A is a surplus with respect to $F \setminus A$, hence $F \setminus A$ has a marriage. By the Main Lemma, there is a K critical in $F \setminus A$ such that $F \setminus (A \cup \{a\})$ has a marriage for each $a \notin F(K)$.

If $a \in X \setminus A$, then $F \setminus (A \cup \{a\})$ has no marriage, and therefore $a \in F(K)$. This implies $X \subseteq F(K) \cup A$.

Theorem 3.11. If A is a finite maximal surplus of X with respect to F and $B \subseteq X$ is a surplus with respect to F, then $|B| \leq |A|$.

Proof. By Theorem 3.10, there is a K critical in $F \setminus A$ with $X \subseteq F(K) \cup A$. Since $B \subseteq X \subseteq F(K) \cup A$, we get $|B| \leq |A|$ by Fact 4.

The restriction that A is finite is unessential, since we can prove Theorem 3.11 for infinite sets A with help of Theorem II.2.10 which is more general then Fact 4. But we will only use this special case.

Lemma 3.12. If M is maximal matchable in F, then there is a set K maximal critical in F such that

$$M = K \cup \{i : F(i) \not\subseteq F(K)\}.$$

Proof. By the Main Lemma, there is a maximal critical K in $F \upharpoonright M$ such that $F \upharpoonright M \setminus \{a\}$ has a marriage for each $a \notin F(K)$.

If $i \notin M$, then $F \upharpoonright M \cup \{i\}$ has no marriage. Hence $F(i) \subseteq F(K)$ and so $i \notin K \cup \{i : F(i) \not\subseteq F(K)\}$. For the proof of the converse inclusion let $i \notin K \cup \{i : F(i) \not\subseteq F(K)\}$. Then $i \notin K$ and $F(i) \subseteq F(K)$, and $F \upharpoonright K \cup \{i\}$ has no marriage. Hence $F \upharpoonright M \cup \{i\}$ has no marriage, which implies $i \notin M$.

It remains to show that K is maximal critical in F. Since $F(\mathrm{dom}\, F \setminus M) \subseteq F(K)$, this is obvious.

By Theorem 1.9, every Hall family has a maximal matchable subfamily. But there are families which possess no maximal matchable set, as for example $F = \{(\alpha, \omega) : \alpha \in \omega_1\}$. For this family the cardinal

$$d(F) = \min\{|I \setminus J| : J \text{ is matchable in } F\}$$

is uncountable.

Theorem 3.13. (Aharoni [A1])
If d(F) is countable, then there exists a set M maximal matchable in F.

Proof. Let K be maximal critical in F and take

$$M = K \cup \{i : F(i) \not\subseteq F(K)\}.$$

It is sufficient to show that $F \upharpoonright M$ has a marriage. Since d(F) is countable, there is a countable J such that $(F \upharpoonright M) \setminus J \in \mathcal{M}$. We want to apply Theorem 3.8 to show that $F \upharpoonright M$ has a marriage. So we have to prove that $F \upharpoonright M$ is c-good: Let $L \subseteq M$ be critical in F, and let $i \in M$ with $F(i) \subseteq F(L)$. By Corollary 3.2, $F(L) \subseteq F(K)$. Hence $L \cup \{i\} \subseteq K$ by the definition of M. Therefore $F \upharpoonright L \cup \{i\} \in \mathcal{M}$, and consequently $i \in L$. Now Corollary 3.5 implies that $F \upharpoonright M$ is c-good.

Lemma 3.14. Let M, N be maximal matchable in F, and let f be a marriage of $F \upharpoonright M$. Then there is a marriage g of $F \upharpoonright N$ with $\mathrm{rng}\, f = \mathrm{rng}\, g$.

Proof. By Lemma 3.12, there is a set K maximal critical in F with $M = K \cup \{i : F(i) \not\subseteq F(K)\}$. Since K is critical, $f[K] = F(K)$. Therefore $f \upharpoonright (M \setminus K)$ is a marriage of $F \upharpoonright (M \setminus K) \setminus F(K)$. Again by Lemma 3.12, there is a set L maximal critical in F with $N = L \cup \{i : F(i) \not\subseteq F(L)\}$. By Corollary 3.2, $F(K) = F(L)$. This implies $N \setminus L = M \setminus K$, and therefore $f \upharpoonright (M \setminus K)$ is a marriage of $F \upharpoonright (N \setminus L) \setminus F(L)$. Let h be a marriage of $F \upharpoonright L$. Since L is critical, $h[L] = F(L) = F(K) = f[K]$. Now $g = h \cup f \upharpoonright (M \setminus K)$ has the desired properties.

Lemma 3.15. (König's Duality Theorem)
Let $\Gamma = (A,B,E)$ be a bipartite graph which possesses a maximal matchable set $M \subseteq A$. Then there is a matching f of M and an A-B-separating set S consisting of exactly one vertex of each edge of f. Moreover, if g is a matching of an arbitrary maximal matchable set $N \subseteq A$, then S consists again of exactly one vertex of each edge of g.

Proof. By Lemma 3.12, there are sets K, L maximal critical in F such that

$M = K \cup \{i : F(i) \not\subseteq F(K)\}$ and

$N = L \cup \{i : F(i) \not\subseteq F(L)\}$.

By Corollary 3.2, $F(K) = F(L)$ and so $M \setminus K = N \setminus L$. Let f be a matching of M and put $S = (M \setminus K) \cup F(K)$. Then S has the desired properties.

§4. Inductive definitions of critical sets

We already know two ways how to obtain critical families from critical families. The first one follows from Fact 1:

If $\mathcal{K}$ is a chain of critical sets in F with $\operatorname{dom} F = \bigcup \mathcal{K}$, then F is critical.

The second method follows from Corollary 3.3:

If $i \in \operatorname{dom} F$ and $a \in F(i)$ such that $F(i) \subseteq F(\operatorname{dom} F \setminus \{i\}) \cup \{a\}$ and $F \setminus \{i\} \setminus \{a\}$ is critical, then F is critical.

We can use this to define a hierarchy $\mathcal{C}_\alpha$ of critical families by recursion on $\alpha \in \mathrm{On}$.

Definition. If $\alpha = 0$, then $\mathscr{C}_\alpha = \{\emptyset\}$. If α is a limit ordinal, then $F \in \mathscr{C}_\alpha$ if there is a chain $\mathscr{K}$ such that $\bigcup \mathscr{K} = \operatorname{dom} F$ and, for each $K \in \mathscr{K}$, there is a $\beta < \alpha$ with $F \restriction K \in \mathscr{C}_\beta$. If $\alpha = \beta + 1$, then $F \in \mathscr{C}_\alpha$ if $F \in \mathscr{C}_\beta$ or there are an $i \in \operatorname{dom} F$ and an $a \in F(i)$ such that $F(i) \subseteq F(\operatorname{dom} F \setminus \{i\}) \cup \{a\}$ and $F \setminus \{i\} \setminus \{a\} \in \mathscr{C}_\beta$.

By Fact 1 and Corollary 3.3, we get by induction on $\alpha \in On$:

Lemma 4.1. If $F \in \mathscr{C}_\alpha$, then F is critical.

Example. Let $\alpha \in On$, and let $F_\alpha = (\beta + 1 : \beta \in \alpha)$. We show by induction on α that $F_\alpha \in \mathscr{C}_\alpha$.

If $\alpha = 0$, then $F_\alpha = \emptyset \in \mathscr{C}_\alpha$. If α is a limit ordinal, then $\mathscr{K} = \alpha = \{\beta : \beta < \alpha\}$ is a chain with $\bigcup \mathscr{K} = \alpha$. $F_\alpha \restriction \gamma = F_\gamma$ for each $\gamma \in \mathscr{K}$. Hence, by the induction hypothesis, $F_\gamma \in \mathscr{C}_\gamma$ for each $\gamma < \alpha$, and therefore $F_\alpha \in \mathscr{C}_\alpha$ by the definition of $\mathscr{C}_\alpha$.

If $\alpha = \beta + 1$, let $i = \beta$, and let $a = \beta$. Then $F_\alpha \setminus \{i\} \setminus \{a\} = F_\beta$ and $F_\alpha(i) \subseteq F_\alpha(\alpha \setminus \{i\}) \cup \{a\}$. By the induction hypothesis, $F_\beta \in \mathscr{C}_\beta$, and so $F_\alpha \in \mathscr{C}_\alpha$ by the definition of $\mathscr{C}_\alpha$.

We want to show that $\mathscr{C} = \bigcup \{\mathscr{C}_\alpha : \alpha \in On\}$ is the class of all critical families. For this reason we call a class $\mathscr{A}$ of families closed under union of chains if $F \in \mathscr{A}$ whenever there is a chain $\mathscr{K}$ such that $\bigcup \mathscr{K} = \operatorname{dom} F$ and $F \restriction K \in \mathscr{A}$ for every $K \in \mathscr{K}$.
$\mathscr{A}$ is said to be closed under joining pairs if $F \in \mathscr{A}$ whenever there is an $i \in \operatorname{dom} F$ and an $a \in F(i)$ such that $F(i) \subseteq F(\operatorname{dom} F \setminus \{i\}) \cup \{a\}$ and $F \setminus \{i\} \setminus \{a\} \in \mathscr{A}$.

By the definition of the classes $\mathscr{C}_\alpha$, $\mathscr{C}$ is closed under union of chains and joining pairs.

Let F be a critical family. We want to show that $F \in \mathscr{C}$. For this reason we call $J \subseteq \operatorname{dom} F$ a $\mathscr{C}$-set in F if $F \restriction J \in \mathscr{C}$.

Lemma 4.2.
(i) For any $\mathscr{C}$-set K in F, there is a maximal $\mathscr{C}$-set $L \supseteq K$ in F.
(ii) There is a maximal $\mathscr{C}$-set K in F.

Proof. $\mathscr{C}$ is closed under union of chains. Hence, by Zorn's Lemma, (i) holds. (ii) follows from (i), since $\emptyset$ is a $\mathscr{C}$-set in F.

Let K be a maximal $\mathscr{C}$-set in the critical family F. We want to show that $K = \operatorname{dom} F$. To prove this, we need the following lemma.

Lemma 4.3. If K is a maximal $\mathscr{C}$-set in F and $a \in F(i) \setminus F(K)$, then there is a maximal $\mathscr{C}$-set $L \supseteq K$ in $(F \setminus \{i\}) \setminus \{a\}$ such that $F(i) \not\subseteq F(L) \cup \{a\}$.

Proof. Since $a \notin F(K)$, it follows that $i \notin K$. Thus K is a $\mathscr{C}$-set in $F \setminus \{i\} \setminus \{a\}$. By Lemma 4.2, there is a maximal $\mathscr{C}$-set $L \supseteq K$ in $F \setminus \{i\} \setminus \{a\}$.

Suppose that $F(i) \subseteq F(L) \cup \{a\}$. Since $F \restriction L \setminus \{i\} \setminus \{a\} \in \mathscr{C}$ and $\mathscr{C}$ is closed under joining pairs, $F \restriction L \cup \{i\} \in \mathscr{C}$. Now $K \subseteq L \cup \{i\}$ is a maximal $\mathscr{C}$-set in F, therefore $L \cup \{i\} = K$ and so $a \in F(i) \subseteq F(K)$. This contradiction shows that $F(i) \not\subseteq F(L) \cup \{a\}$.

Now we can prove

Theorem 4.4. If F is critical, then $F \in \mathscr{C}$.

Proof. Let f be a marriage of F and K_0 be a maximal $\mathscr{C}$-set in F. We want to show that $\operatorname{dom} F = K_0$ and suppose for contradiction that there is an $i_0 \in \operatorname{dom} F \setminus K_0$. We will choose pairwise different i_n, $n \in \omega$, with $f(i_{n+1}) \in F(i_n)$. Then

$$g = (f \setminus \{(i_n, f(i_n)) : n \in \omega\}) \cup \{(i_n, f(i_{n+1})) : n \in \omega\}$$

is a marriage of F with $f(i_0) \notin \operatorname{rng} g$, which contradicts the fact that F is critical.

Suppose that we have already chosen $i_0, \ldots, i_n$ and a maximal $\mathscr{C}$-set K_n in $F_n := F \setminus f \restriction \{i_k : k < n\}$ such that $i_n \notin K_n$. $F_n \restriction K_n \in \mathscr{C}$, so, by Lemma 4.1, K_n is critical in F_n, and consequently $f[K_n] = F_n(K_n)$. Hence $f(i_n) \in F_n(i_n) \setminus F_n(K_n)$. By Lemma 4.3, there is a maximal $\mathscr{C}$-set $K_{n+1} \supseteq K_n$ in F_{n+1} with $F_n(i_n) \not\subseteq F_n(K_{n+1}) \cup \{f(i_n)\}$. Since F is critical, there is an $i_{n+1} \in \operatorname{dom} F \setminus K_{n+1}$ such that $f(i_{n+1}) \in F_n(i_n) \setminus (F_n(K_{n+1}) \cup \{f(i_n)\}) =$
$= F(i_n) \setminus (F(K_{n+1}) \cup \{f(i_k) : k \leq n\})$. This completes the definition of the sequence $(i_n : n \in \omega)$, and the theorem is proved.

Corollary 4.5.
The following propositions are equivalent:
(1) F is critical.
(2) There is an α with $F \in \mathcal{C}_\alpha$.

Thus the class of all critical families is the smallest class which is closed under union of chains and joining pairs.

Definition. We define a second hierarchy of families by recursion on $\alpha \in On$.

If $\alpha = 0$, then let $\vartheta_\alpha = \{\emptyset\}$. If α is a limit ordinal, then $F \in \vartheta_\alpha$ if there is a continuous chain $(K_\gamma : \gamma \leq \delta)$ with $K_\delta = \operatorname{dom} F$ such that, for every $\gamma < \delta$, there is a $\beta < \alpha$ with $(F \upharpoonright K_{\gamma+1} \setminus K_\gamma) \setminus F(K_\gamma) \in \vartheta_\beta$. If $\alpha = \beta + 1$, then $F \in \vartheta_\alpha$ if $F \in \vartheta_\beta$ or there are an $i \in \operatorname{dom} F$ and an $a \in F(i)$ such that $F(i) \subseteq F(\operatorname{dom} F \setminus \{i\}) \cup \{a\}$ and $F \setminus \{i\} \setminus \{a\} \in \vartheta_\beta$.

Example. Again let $F_\alpha = (\beta + 1 : \beta \in \alpha)$. Then $(\beta : \beta \leq \alpha)$ is a continuous chain and, for every $\beta < \alpha$, we have $((F_\alpha \upharpoonright \beta + 1) \setminus \beta) \setminus F_\alpha(\{\delta : \delta < \beta\}) = \{(\beta, \{\beta\})\} \in \vartheta_1$. Thus $F_\alpha \in \vartheta_\omega$.

Example. For each $\alpha \in On$, we define a family $F_\alpha \in \vartheta_\alpha \setminus \cup \{\vartheta_\gamma : \gamma < \alpha\}$ as follows: Let $F_0 = \emptyset$ and assume that we have defined F_β for each $\beta < \alpha$. If $\alpha = \beta + 1$, take $i \notin \operatorname{dom} F_\beta$, $a \notin F_\beta(\operatorname{dom} F_\beta)$, and define $\operatorname{dom} F_\alpha = \operatorname{dom} F_\beta \cup \{i\}$, $F_\alpha(i) = \{a\}$, and $F_\alpha(j) = F_\beta(j) \cup \{a\}$ for each $j \in \operatorname{dom} F_\beta$. If α is a limit ordinal, then choose disjoint copies F'_β of the families F_β, $\beta < \alpha$ ($\operatorname{dom} F'_\beta \cap \operatorname{dom} F'_\gamma = \emptyset$ and $F'(\operatorname{dom} F'_\beta) \cap F'(\operatorname{dom} F'_\gamma) = \emptyset$, whenever $\beta < \gamma < \alpha$), and take $F_\alpha = \cup \{F'_\beta : \beta < \alpha\}$.

Define $\vartheta = \cup \{\vartheta_\alpha : \alpha \in On\}$. Again we want to show that ϑ is the class of all critical families. The fact that every $F \in \vartheta$ is critical follows from the following lemma.

Lemma 4.6. Let $(K_\gamma : \gamma \leq \rho)$ be a continuous chain of subsets of $\operatorname{dom} F$. If, for any $\gamma < \rho$, $K_{\gamma+1} \setminus K_\gamma$ is critical in $F \setminus F(K_\gamma)$, then K_ρ is critical in F.

Proof. We prove the lemma by induction on ρ.
If $\rho = 0$, then $K_\rho = \emptyset$ is critical in F. Let ρ be a successor ordinal $\gamma + 1$. Then K_γ is critical in F by the induction hypothesis, hence K_ρ

is critical in F by Fact 2. Finally let ρ be a limit ordinal. Then each K_γ, $\gamma<\rho$, is critical in F by the induction hypothesis. Hence $K_\gamma = \cup\{K_\gamma : \gamma<\rho\}$ is critical in F by Fact 1.

By Lemma 4.6 and Corollary 3.3, we get by induction on α:

Corollary 4.7. If $F\in\vartheta_\alpha$, then F is critical.

If F is a critical family, then we have shown that $F\in\mathscr{C}_\alpha$ for some ordinal α. To prove this, we used the fact that $\mathscr{C}$ is closed under union of chains. It is not immediately to be seen that $\vartheta = \cup\{\vartheta_\alpha : \alpha\in \mathrm{Ord}\}$ is closed under union of chains, but it is easier to show that $\mathscr{C}_\alpha \subseteq \vartheta_\alpha$ for each $\alpha\in\mathrm{On}$, and this proves, by Theorem 4.4, that every critical family is an element of ϑ.

Lemma 4.8. If $F\in\mathscr{M}$, $F\restriction K\in\mathscr{C}_\beta$, and L is critical in F, then $(F\restriction K\setminus L)\setminus F(L)\in\mathscr{C}_\beta$.

Proof. We prove the lemma by induction on β. If $\beta = 0$, then $F\restriction K = \emptyset$ and so $(F\restriction K\setminus L)\setminus F(L) = \emptyset\in\mathscr{C}_\beta$. If β is a limit ordinal, then, by the definition of $\mathscr{C}_\beta$ and the induction hypothesis, $(F\restriction K\setminus L)\setminus F(L)\in\mathscr{C}_\beta$.

Let β be a successor ordinal $\gamma+1$. If $F\restriction K\in\mathscr{C}_\gamma$, then, by the induction hypothesis, $(F\restriction K\setminus L)\setminus F(L)\in\mathscr{C}_\gamma\subseteq\mathscr{C}_\beta$. Otherwise there are an $i\in K$ and an $a\in F(i)$ such that $F(i)\subseteq F(K\setminus\{i\})\cup\{a\}$ and $(F\restriction K\setminus\{i\})\setminus\{a\}\in\mathscr{C}_\gamma$.

Let f be a marriage of F. Since $F\restriction K$ and $F\restriction K\setminus\{i\}\setminus\{a\}$ are critical, we can assume that $f(i) = a$. Thus $(F\setminus\{i\})\setminus\{a\}\in\mathscr{M}$ and $L\setminus\{i\}$ is critical in $(F\setminus\{i\})\setminus\{a\}$. Since L is critical in F, $i\in L$ if and only if $a = f(i)\in F(L)$, and further $F(L\setminus\{i\})\cup\{a\} = F(L)\cup\{a\}$. So, by the induction hypothesis, we get $(F\restriction K\setminus(L\cup\{i\}))\setminus(F(L)\cup\{a\})\in\mathscr{C}_\gamma$. If $i\in L$, then $(F\restriction K\setminus L)\setminus F(L) = (F\restriction K\setminus(L\cup\{i\}))\setminus(F(L)\cup\{a\})\in\mathscr{C}_\gamma\subseteq\mathscr{C}_\beta$. Otherwise $a\in F(i)\setminus F(L)$ and $F(i)\setminus F(L)\subseteq(F(K\setminus(L\cup\{i\}))\setminus F(L))\cup\{a\}$, and so $(F\restriction K\setminus L)\setminus F(L)\in\mathscr{C}_\beta$ by the definition of $\mathscr{C}_\beta$.

Theorem 4.9. If $F\in\mathscr{C}_\beta$, then $F\in\vartheta_\beta$.

Proof. We prove the theorem by induction on β.

If $\beta = 0$, then $\mathscr{C}_\beta = \{\emptyset\} = \vartheta_\beta$. If β is a successor ordinal $\gamma+1$, then, by the induction hypothesis and the definitions of $\mathscr{C}_\beta$ and ϑ_β, we have $\mathscr{C}_\beta \subseteq \vartheta_\beta$. Let β be a limit ordinal and $F \in \mathscr{C}_\beta$. Then there is a chain $(K_\gamma : \gamma < \delta)$ such that $\cup\{K_\gamma : \gamma < \delta\} = \operatorname{dom} F$ and, for each $\gamma < \delta$, there is an $\alpha_\gamma < \beta$ with $F \upharpoonright K_\gamma \in \mathscr{C}_{\alpha_\gamma}$. If $L_\gamma = \cup\{K_\alpha : \alpha < \gamma\}$ for $\gamma \leq \delta$, then $(L_\gamma : \gamma \leq \delta)$ is a continuous chain with $L_\delta = \operatorname{dom} F$. Since $F \in \mathscr{C}_\beta$, F is critical and so $F \in \mathscr{M}$. If $\alpha < \gamma \leq \delta$, then $F \upharpoonright K_\alpha \in \mathscr{C}_\beta$ and so it is critical. Further $F \upharpoonright L_\gamma \in \mathscr{M}$. Therefore $F \upharpoonright L_\gamma$ is critical as a union of critical subfamilies. By Lemma 4.8, $(F \upharpoonright L_{\gamma+1} \setminus L_\gamma) \setminus F(L_\gamma) = (F \upharpoonright K_\gamma \setminus L_\gamma) \setminus F(L\ ) \in \mathscr{C}_{\alpha_\gamma}$ and so, by the induction hypothesis, $(F \upharpoonright L_{\gamma+1} \setminus L_\gamma) \setminus F(L_\gamma) \in \vartheta_{\alpha_\gamma}$. Hence $F \in \vartheta_\beta$ by the definition of ϑ_β.

<u>Corollary 4.10</u>. A family F is critical if and only if there is an ordinal β with $F \in \vartheta_\beta$.

Let us say that a class $\mathscr{A}$ is closed under union of ladders if $F \in \mathscr{A}$ whenever there is a continuous chain $(K_\alpha : \alpha \leq \rho)$ such that $\operatorname{dom} F = K_\rho$ and $(F \upharpoonright K_{\alpha+1} \setminus K_\alpha) \setminus F(K_\alpha) \in \mathscr{A}$ for every $\alpha < \rho$.

Then the class of all critical families in the smallest class which is closed under union of ladders and under joining pairs.

Finally we define a third hierarchy of families.
Suppose that for all $\beta < \alpha$ we have already defined $\mathscr{F}_\beta$.

We call a family F <u>α-finite</u> if there are $\beta < \alpha$, $L \subset\subset \operatorname{dom} F$, and $A \subset\subset F(L)$ such that

1. $F(L) \subseteq F(\operatorname{dom} F \setminus L) \cup A$,
2. $(F(i) \cap A : i \in L)$ is critical, and
3. $(F \setminus L) \setminus A \in \mathscr{F}_\beta$.

A set $\mathscr{K}$ is said to be a <u>directed system</u> if, for any $L, K \in \mathscr{K}$, there exists an $M \in \mathscr{K}$ with $L \cup K \subseteq M$.

Then let $F \in \mathscr{F}_\alpha$ whenever there is a directed system $\mathscr{K}$ such that $\cup\mathscr{K} = \operatorname{dom} F$ and $F \upharpoonright K$ is α-finite for every $K \in \mathscr{K}$.

If $\alpha = 0$, then $\emptyset$ is the only directed system with the desired property. Hence $\mathscr{F}_0 = \{\emptyset\}$.

So, by definition, we get the equivalence of

(i) F is 1-finite.
(ii) F is a finite critical Hall family.

Claim. $F \in \mathcal{F}_1$ if and only if F is a critical Hall family.

Proof. If $F \in \mathcal{F}_1$, then there is a directed system $\mathcal{K}$ such that $\bigcup\mathcal{K} = \operatorname{dom} F$ and $F \restriction K$ is 1-finite for each $K \in \mathcal{K}$. Thus, for each $K \in \mathcal{K}$, $F \restriction K$ is a finite critical Hall family. Therefore it is sufficient to show that F has a marriage. Let $L \subset\subset \operatorname{dom} F$. Since $\mathcal{K}$ is a directed system with $\bigcup\mathcal{K} = \operatorname{dom} F$, there is a $K \in \mathcal{K}$ such that $L \subseteq K$, which implies that $F \restriction L$ has a marriage. Therefore, by the Compactness Theorem 1.4, F has a marriage.

Conversely let F be a critical Hall family. We put $\mathcal{K} = \{K : K \subset\subset \operatorname{dom} F$ and $F \restriction K$ is critical$\}$. Then every $K \in \mathcal{K}$ is 1-finite. Since F has a marriage, $\mathcal{K}$ is a directed system. So it remains to show that $\bigcup\mathcal{K} = \operatorname{dom} F$. Let $i \in \operatorname{dom} F$. By Theorem 1.6, there is a set $K \subset\subset \operatorname{dom} F$ critical in F with $F(i) \subseteq F(K)$. Now F has a marriage, and so $i \in K$.

Again we want to show that $\mathcal{F} = \bigcup\{\mathcal{F}_\alpha : \alpha \in On\}$ is the class of all critical families. Since $\mathcal{F}$ is closed under union of chains and joining pairs, we get

Lemma 4.11. If F is critical, then $F \in \mathcal{F}_\alpha$ for some ordinal α.

To prove that every $F \in \mathcal{F}$ is critical, we prove two lemmata.

Lemma 4.12. Let $L \subset\subset \operatorname{dom} F$ and $A \subset\subset F(L)$ such that $F(L) \subseteq F(\operatorname{dom} F \setminus L) \cup A$. If $(F(i) \cap A : i \in L)$ and $(F \setminus L) \setminus A$ are critical, then F is critical.

Proof. Let f be a marriage of $F \setminus L \setminus A$ and g a marriage of $(F(i) \cap A : i \in L)$. Then $f \cup g$ is a marriage of F.

Now let g be an arbitrary marriage of F. We want to show that $\operatorname{rng} g = F(\operatorname{dom} F)$. Let $K = \operatorname{dom} F \setminus L$ and $B = F(\operatorname{dom} F) \setminus g[K]$. Then $|B| \leq |A|$ by Fact 4. Since $A \subset\subset F(L)$ and $(F(i) \cap A : i \in L)$ is critical, we have $|A| = |L|$. Finally $g[L] \subseteq B$, and therefore $|g[L]| \leq |B| \leq |A| \leq |L|$. So $g[L] = B = F(\operatorname{dom} F) \setminus g[K]$, which implies $\operatorname{rng} g = F(\operatorname{dom} F)$.

Lemma 4.13. If $\mathcal{K}$ is a directed system of critical sets in F with $\bigcup\mathcal{K} = \mathrm{dom}\,F$, then F is critical.

Proof. A critical set L in F is called admissible if $L \cup K$ is critical in F for every $K \in \mathcal{K}$. Clearly $\emptyset$ is admissible.

Claim 1. There is a maximal admissible set.

Let $\mathcal{B}$ be a chain of admissible sets, and let $K \in \mathcal{K}$. Then $\{L \cup K : L \in \mathcal{B}\}$ is a chain of critical sets in F. Hence $\bigcup\mathcal{B} \cup K$ is critical in F by Fact 1. Therefore Claim 1 follows from Zorn's Lemma.

Claim 2. If L is admissible and $N \in \mathcal{K}$, then $L \cup N$ is admissible.

Let $K \in \mathcal{K}$. Then there is a set $M \in \mathcal{K}$ with $N \cup K \subseteq M$. L is admissible, so $L \cup M$ is critical and $F \upharpoonright L \cup N \cup K$ has a marriage. Since L, K and N are critical in F, $L \cup N \cup K$ is critical in F. This proves Claim 2.

By Claim 1, there is a maximal admissible set L_0. Claim 2 implies that $\mathrm{dom}\,F = \bigcup\mathcal{K} \subseteq L_0$, which proves that F is critical.

Theorem 4.14.
The following propositions are equivalent:
(a) F is critical.
(b) $F \in \mathcal{F}_\alpha$ for some ordinal α.

Proof. (a) implies (b) by Lemma 4.11. The converse follows by induction on α from Lemma 4.12 and Lemma 4.13.

§5. Queues

In this section we want to give a criterion similar to Nash-Williams' criterion in [N3]. Let $\mathfrak{Z}$ denote the set the elements of which are the integers and two further "numbers" ∞ and $-\infty$. The size $\|X\|$ of a set X is defined to be its cardinality $|X|$ if X is finite, and to be ∞ if X is infinite. For each $n \in \mathbb{Z}$ we fix the following additional rules:

$$n + \infty = \infty + n = \infty + \infty = \infty,$$
$$-\infty + n = -\infty + \infty = -\infty, \text{ and}$$
$$-\infty < n < \infty.$$

If α is a limit ordinal, $\gamma < \alpha$, and $(q_\delta : \delta < \alpha)$ a sequence of elements of

$\beth$, then it is not difficult to show that

$$\sup_{\beta<\alpha} \inf_{\beta<\delta<\alpha} q_\delta = \sup_{\gamma<\beta<\alpha} \inf_{\beta<\delta<\alpha} q_\delta .$$

Let $F = (F(i) \mid i \in I)$ be a family, and let α be an ordinal. We define a <u>queue in F</u> to be an injective function $k : \alpha \to I$. If k is surjective, we call k a <u>queue of F</u>. α is said to be the <u>length</u> of k, denoted by $\mathrm{lh}(k)$.

By transfinite recursion on the length α of k we define <u>$p_F(k) \in \beth$</u> as follows:

If $\alpha = 0$, then $p_F(k) = 0$.

If α is a successor ordinal $\beta + 1$, then

$$p_F(k) = p_F(k \restriction \beta) + \| F(k(\beta)) \setminus F(k[\beta]) \| - 1.$$

If α is a limit ordinal, then

$$p_F(k) = \sup_{\gamma<\alpha} \inf_{\gamma<\delta<\alpha} p_F(k \restriction \delta)$$

Roughly speaking, $p_F(k)$ is the largest number of women of $F(\mathrm{rng}\, k)$ we could hope to leave unmarried after we have married $\mathrm{rng}\, k$. This is made precise by the following lemma.

<u>Lemma 5.1</u>. Let k be a queue in F. If f is a marriage of F, then

$$\| F(\mathrm{rng}\, k) \setminus f(\mathrm{rng}\, k) \| \leq p_F(k).$$

<u>Proof</u>. Let α be the length of k, $i_\beta = k(\beta)$ for $\beta < \alpha$, and $I_\beta = k[\beta]$ for $\beta \leq \alpha$. Then $I_\alpha = \mathrm{rng}\, k$. By transfinite induction on $\beta \leq \alpha$ we show that

$$\| F(I_\beta) \setminus f[I_\beta] \| \leq p_F(k \restriction \beta).$$

If $\beta = 0$, then $I_\beta = \emptyset$ and $p_F(k \restriction \beta) = 0$, and the assertion is true.

If $\beta = \gamma + 1$, then $\| F(I_\gamma) \setminus f[I_\gamma] \| \leq p_F(k \restriction \gamma)$ by the induction hypothesis. Hence

$$\begin{aligned} p_F(k \restriction \beta) &= p_F(k \restriction \gamma) + \| F(i_\gamma) \setminus F(I_\gamma) \| - 1 \\ &\geq \| F(I_\gamma) \setminus f[I_\gamma] \| + \| F(i_\gamma) \setminus F(I_\gamma) \| - 1 \\ &= \| F(I_\beta) \setminus f[I_\gamma] \| - 1 \\ &= \| F(I_\beta) \setminus f[I_\beta] \| . \end{aligned}$$

Let β be a limit ordinal. Then, for every $\gamma<\beta$,

$$\|F(I_\gamma)\setminus f[I_\gamma]\| \leq p_F(k \upharpoonright \gamma)$$

by the induction hypothesis. Since $\cup\{F(I_\gamma):\gamma<\beta\} = F(I_\beta)$, we have

$$\|F(I_\beta)\setminus f[I_\beta]\| = \sup_{\gamma<\beta} \|F(I_\gamma)\setminus f[I_\beta]\|$$

Hence

$$\begin{aligned} p_F(k \upharpoonright \beta) &= \sup_{\gamma<\beta} \inf_{\gamma<\delta<\beta} p_F(k \upharpoonright \delta) \\ &\geq \sup_{\gamma<\beta} \inf_{\gamma<\delta<\beta} \|F(I_\delta)\setminus f[I_\delta]\| \\ &\geq \sup_{\gamma<\beta} \|F(I_\gamma)\setminus f[I_\beta]\| \\ &= \|F(I_\beta)\setminus f[I_\beta]\|. \end{aligned}$$

We will say that a family F is <u>p-good</u> if $p_F(k)\geq 0$ for every queue k in F. Otherwise we call F <u>p-bad</u>. If F has a marriage, then F is p-good by Lemma 5.1.

To prove that every countable p-good family has a marriage, we need some lemmata similar to Facts 1-4 of Chapter 3.

We call a queue k <u>p-critical</u> in F if $p_F(k)\leq 0$.

<u>Lemma 5.2</u>. If $\mathcal{K}$ is a chain of p-critical queues in F, then $\cup\mathcal{K}$ is p-critical in F.

<u>Proof</u>. Let $l := \cup\mathcal{K}$. If $\mathcal{K} = \emptyset$, then $l = \emptyset$ and so $p_F(l) = 0$. Assume that $\mathcal{K} \neq \emptyset$ and let $\Gamma = \{lh(k) : k\in\mathcal{K}\}$. Then $lh(l) = \cup\Gamma$. If $lh(l)\in\Gamma$, then there is a $k\in\mathcal{K}$ with $l = k$, which implies $p_F(l) = p_F(k)\leq 0$. Otherwise, for each $\gamma< lh(l)$, there is a $k\in\mathcal{K}$ such that $\gamma< lh(k) < lh(l)$. Since $p_F(k)\leq 0$ for every $k\in\mathcal{K}$, $p_F(l)\leq 0$ by the definition of p_F.

<u>Corollary 5.3</u>. There is a maximal p-critical queue in F.

<u>Definition</u>. If k is a queue in F, we define $F\setminus k := (F\setminus \mathrm{rng}\,k)\setminus F(\mathrm{rng}\,k)$. Let l be a queue in $F\setminus k$. Then $k*l$ is defined by

$$(k*l)(\gamma) = k(\gamma) \text{ for } \gamma< lh(k),$$

$$(k*l)(\gamma) = l(\alpha) \text{ for } \alpha< lh(l) \text{ and } \gamma = lh(k)+\alpha.$$

Lemma 5.4. If k is a queue in F and l a queue in $F\setminus k$, then
$p_F(k*l) = p_F(k)+p_{F\setminus k}(l)$.

Proof. We prove the lemma by induction on lh(l). If lh(l) = 0, then $k*l = k$ and $p_{F\setminus k}(l) = 0$. Assume that $lh(l) = \gamma+1$. Then

$$p_F(k*l) = p_F(k*(l\restriction\gamma)) + \|F(l(\gamma))\setminus F(\mathrm{rng}\,k\cup l[\gamma])\| - 1$$
$$= p_F(k)+p_{F\setminus k}(l\restriction\gamma) + \|(F\setminus k)(l(\gamma))\setminus(F\setminus k)(l[\gamma])\| - 1$$
$$= p_F(k)+p_{F\setminus k}(l)$$

Now let lh(l) be a limit ordinal β, and let $\alpha = lh(k)$. Then

$$p_F(k*l) = \sup_{\gamma<\alpha+\beta}\ \inf_{\gamma<\delta<\alpha+\beta}\ p_F((k*l)\restriction\delta)$$
$$= \sup_{\alpha<\gamma<\alpha+\beta}\ \inf_{\gamma<\delta<\alpha+\beta}\ p_F((k*l)\restriction\delta)$$
$$= \sup_{\gamma<\beta}\ \inf_{\gamma<\delta<\beta}\ p_F(k*(l\restriction\delta))$$
$$= \sup_{\gamma<\beta}\ \inf_{\gamma<\delta<\beta}\ (p_F(k)+p_{F\setminus k}(l\restriction\delta))$$
$$= p_F(k)+p_{F\setminus k}(l).$$

Corollary 5.5. If k is a maximal p-critical queue in F, then $\emptyset$ is a maximal p-critical queue in $F\setminus k$.

Proof. Suppose that l is a p-critical queue in $F\setminus k$. By Lemma 5.4, $k*l$ is p-critical in F. Now k is maximal p-critical in F, so lh(l) = 0.

Lemma 5.6. If k is a queue in F and $A\subseteq F(\mathrm{dom}\,F)$ is finite, then

$$p_F(k) = p_{F\setminus A}(k) + |F(\mathrm{rng}\,k)\cap A|$$

Proof. The proof is by induction on lh(k).
If lh(k) = 0, then $p_F(k) = p_{F\setminus A}(k) = |F(\mathrm{rng}\,k)\cap A| = 0$. Assume $lh(k) = \beta+1$. Then

$$p_F(k) = p_F(k\restriction\beta) + \|F(k(\beta))\setminus F(k[\beta])\| - 1$$
$$= p_{F\setminus A}(k\restriction\beta) + |F(k[\beta])\cap A| +$$
$$+ |(F(k(\beta))\cap A)\setminus F(k[\beta])| + \|(F(k(\beta))\setminus(F(k[\beta])\cup A)\| - 1$$
$$= p_{F\setminus A}(k) + |F(\mathrm{rng}\,k)\cap A|.$$

Let $lh(k) = \alpha$ be a limit ordinal. Since A is finite, there is a $\gamma<\alpha$ with $F(k[\gamma]) \cap A = F(\mathrm{rng}\, k) \cap A$. Then

$$\begin{aligned} p_F(k) &= \sup_{\beta<\alpha} \inf_{\beta<\delta<\alpha} p_F(k \restriction \delta) \\ &= \sup_{\gamma<\beta<\alpha} \inf_{\beta<\delta<\alpha} p_F(k \restriction \delta) \\ &= \sup_{\gamma<\beta<\alpha} \inf_{\beta<\delta<\alpha} (p_{F\setminus A}(k \restriction \delta) + |F(k[\delta]) \cap A|) \\ &= \sup_{\gamma<\beta<\alpha} \inf_{\beta<\delta<\alpha} (|F(\mathrm{rng}\, k) \cap A| + p_{F\setminus A}(k \restriction \delta)) \\ &= |F(\mathrm{rng}\, k) \cap A| + p_{F\setminus A}(k). \end{aligned}$$

Corollary 5.7. If k is maximal p-critical in F and $a \notin F(\mathrm{rng}\, k)$, then $(F\setminus k)\setminus\{a\}$ is p-good.

Proof. Let l be a queue in $(F\setminus k)\setminus\{a\}$ with $lh(l)>0$. Then $p_{F\setminus k}(l)>0$ by Corollary 5.5. Hence $p_{(F\setminus k)\setminus\{a\}}(l)\geq 0$ by Lemma 5.6.

Lemma 5.8. Let F be p-good, and let $i \in \mathrm{dom}\, F$. Then there is a p-critical k in $F\setminus\{i\}$ and an $a \in F(i)\setminus F(\mathrm{rng}\, k)$ such that $((F\setminus k)\setminus\{i\})\setminus\{a\}$ is p-good.

Proof. Let k be maximal p-critical in $F\setminus\{i\}$, and let $l = k \cup \{(lh(k), i)\}$. Since F is p-good, $p_F(l) \geq 0$. Further $p_F(l) = p_F(k) + \|F(i)\setminus F(\mathrm{rng}\, k)\| - 1$ and k is p-critical in F. Therefore $p_F(k) \leq 0$, and so there must exist an $a \in F(i)\setminus F(\mathrm{rng}\, k)$. Corollary 5.7 now implies that $(F\setminus k)\setminus\{i\}\setminus\{a\}$ is p-good.

Theorem 5.9. Let A be finite, and let J be countable. If $F\setminus A$ is p-good and $F\setminus J$ has a marriage, then $F\setminus A$ has a marriage.

Proof. A finite set $\mathcal{G}$ of p-good families is called a p-good partition of F if:

(i) $\{\mathrm{dom}\, G : G \in \mathcal{G}\}$ is a partition of $\mathrm{dom}\, F$ and

(ii) $\{G(\mathrm{dom}\, G) : G \in \mathcal{G}\}$ is a partition of $F(\mathrm{dom}\, F)$.

A finite matching g is called admissible if there is a p-good partition $\mathcal{G}_g$ of $(F\setminus A)\setminus g$. Since $F\setminus A$ is p-good, $\{F\setminus A\}$ is a p-good partition of $(F\setminus A)\setminus \emptyset$, and therefore $\emptyset$ is admissible.

Claim. If g is admissible and $K \subset\subset \operatorname{dom} F$, then there is an admissible $h \supseteq g$ such that $K \subseteq \operatorname{dom} h$.

We prove the claim by induction on $|K \setminus \operatorname{dom} g|$. If $|K \setminus \operatorname{dom} g| = 0$, let $h = g$. Assume that $|K \setminus \operatorname{dom} g| > 0$ and let $i \in K \setminus \operatorname{dom} g$. By the induction hypothesis, there is an admissible $f \supseteq g$ such that $K \setminus \{i\} \subseteq \operatorname{dom} f$. If $i \in \operatorname{dom} f$, then let $h = f$. Otherwise let $\mathscr{G}_f$ be a p-good partition of $F \setminus A \setminus f$. Then there is a $G \in \mathscr{G}_f$ with $i \in \operatorname{dom} G$. By Lemma 5.8, there is a queue k in G and an $a \in G(i) \setminus G(\operatorname{rng} k)$ such that $G \setminus \{i\} \setminus k \setminus \{a\}$ is p-good. Let $h = f \cup \{(i,a)\}$, and let $\mathscr{G}_h = (\mathscr{G}_f \setminus \{G\}) \cup \{G \restriction \operatorname{rng} k, G \setminus k \setminus \{i\} \setminus \{a\}\}$. Then $\mathscr{G}_h$ is a p-good partition of $F \setminus A \setminus h$, h is admissible, and the claim is proved.

Let f be a marriage of $F \setminus J$, and let $J_n \subset\subset J$ for each $n \in \omega$ such that $\cup\{J_n : n \in \omega\} = J$. Just like in the proof of Theorem 2.8, we choose admissible matchings g_n for each $n \in \omega$. Let $g_0 = \emptyset$ and assume that g_n is already chosen. Then let $K_n = J_n \cup \{i : f(i) \in A \cup \operatorname{rng} g_n\}$ and take $g_{n+1} \supseteq g_n$ with $K_n \subseteq \operatorname{dom} g_{n+1}$, which is possible by the claim. If $g = \cup\{g_n : n \in \omega\}$, then $f \restriction (\operatorname{dom} f \setminus \operatorname{dom} g) \cup g$ is a marriage of $F \setminus A$.

Corollary 5.10.
If F is a countable family, then the following conditions are equivalent:

(1) F has a marriage.

(2) For every queue k in F, $p_F(k) \geq 0$.

Proof. (1) implies (2) by Lemma 5.1. For the converse let $J = \operatorname{dom} F$. Then $F \setminus J = \emptyset$ has a marriage, and $F \in \mathscr{M}$ by (2) and Theorem 5.9.

Theorem 5.11.
For every family F, the following conditions are equivalent:

(1) $p_F(k) \geq 0$ for each queue k in F.

(2) If $F \restriction J$ has a marriage and $i \in \operatorname{dom} F$, then $F \restriction J \cup \{i\}$ has a marriage.

Proof. Assume (1), let $F \restriction J \in \mathscr{M}$, and let $i \in \operatorname{dom} F$. Clearly $F \restriction J \cup \{i\}$ is p-good, since F is p-good, so $F \restriction J \cup \{i\} \in \mathscr{M}$ by Theorem 5.9. Now we prove that (2) implies (1) by induction on $lh(k)$. If $lh(k) = 0$, nothing has to be proved. Let $lh(k)$ be a successor ordinal $\beta + 1$. Since F is c-good, Corollary 3.7 implies that there is an

$a \in F(k(\beta))$ such that $F \setminus \{k(\beta)\} \setminus \{a\}$ is c-good. By the induction hypothesis, $p_{F\setminus\{a\}}(k \upharpoonright \beta) \geq 0$. Hence $p_{F\setminus\{a\}}(k) \geq -1$ by the definition of $p_{F\setminus\{a\}}$, and $p_F(k) \geq 0$ by Lemma 5.6. If $\mathrm{lh}(k)$ is a limit ordinal, then $p_F(k \upharpoonright \delta) \geq 0$ for every $\delta < \mathrm{lh}(k)$ by the induction hypothesis. So $p_F(k) \geq 0$ by the definition of p_F.

Theorem 5.12.
For every family F, the following conditions are equivalent:
(1) F is critical.
(2) (i) $p_F(k) \geq 0$ for each queue k in F, and
(ii) There is a queue k of F with $p_F(k) = 0$.

Proof. If (1) holds, then F has a marriage f and $p_F(k) \geq 0$ for each queue k in F by Lemma 5.1. Let k be a maximal p-critical queue in F. We want to show that $\mathrm{dom}\, F = \mathrm{rng}\, k$ and assume, to get a contradiction, that there is an $i \in \mathrm{dom}\, F \setminus \mathrm{rng}\, k$. Since $p_F(k) \leq 0$, $f[\mathrm{rng}\, k] = F(\mathrm{rng}\, k)$ by Lemma 5.1. Therefore $f(i) \notin F(\mathrm{rng}\, k)$, which implies that $f \upharpoonright (\mathrm{dom}\, F \setminus \mathrm{rng}\, k)$ is a marriage of $F \setminus k$. By Corollary 5.7, $(F \setminus k) \setminus \{f(i)\}$ is p-good. Hence, by Theorem 5.9, $(F \setminus k) \setminus \{f(i)\}$ has a marriage g. Then $h = f \upharpoonright \mathrm{rng}\, k \cup g$ is a marriage of F with $f(i) \notin \mathrm{rng}\, h$, and so F is not critical. Thus $\mathrm{dom}\, F = \mathrm{rng}\, k$ and k is a queue of F with $p_F(k) = 0$.

To prove the converse, let K be maximal critical in F. We want to show that $\mathrm{dom}\, F = K$ and suppose that there is an $i \in \mathrm{dom}\, F \setminus K$. F is p-good by assumption, hence F is c-good by Theorem 5.11. Therefore there exists an $a \in F(i) \setminus F(K)$, and $F \setminus \{a\}$ is c-good by Lemma 3.6. Again by Theorem 5.11, $F \setminus \{a\}$ is p-good. By (ii), there is a queue k of F with $p_F(k) \leq 0$. Now $p_{F\setminus\{a\}}(k) \leq -1$ by Lemma 5.6, which contradicts the fact that $F \setminus \{a\}$ is p-good.

§6. Rank functions

If F is a family, then in this section r denotes a function from $\mathrm{dom}\, F$ into $\gamma \in \mathrm{On}$ and I_α is the set $\{i : r(i) < \alpha\}$ for each $\alpha \leq \gamma$. Hence $I_0 = \emptyset$ and $I_\gamma = \mathrm{dom}\, F$.

r is said to be a rank function of F if, for each $i \in \mathrm{dom}\, F$, there are a finite set A and a set $K \subseteq \{j : r(j) < r(i)\}$ critical in $F \setminus A$ such that $F(i) \subseteq F(K) \cup A$.

If $\gamma = 1$, then $r(i) = 0$ for each $i \in \mathrm{dom}\, F$. Hence $F(i)$ is finite for each $i \in \mathrm{dom}\, F$, and so F is a Hall family. Conversely, if F is a

Hall family, then each function $r: \mathrm{dom}\, F \to \gamma$ is a rank function of F.

An easy exercise is

Lemma 6.1.

(1) If r is a rank function of F and A is a set, then r is a rank function of $F \setminus A$.

(2) If A is a finite set and r is a rank function of $F \setminus A$, then r is a rank function of F.

Proof. For (1), let $i \in \mathrm{dom}\, F$. Then there are a finite set C and a set $K \subseteq \{j : r(j) < r(i)\}$ critical in $F \setminus C$ such that $F(i) \subseteq F(K) \cup C$. By Fact 3, there is a set $L \subseteq K$ critical in $F \setminus (A \cup C) = (F \setminus A) \setminus (C \setminus A)$ such that $F(K) \cup (A \cup C) = F(L) \cup (A \cup C)$. Hence $(F \setminus A)(i) \subseteq (F(K) \cup C) \setminus A = (F(L) \setminus A) \cup (C \setminus A) = (F \setminus A)(L) \cup (C \setminus A)$, which implies that r is a rank function of $F \setminus A$. Now we prove (2). If $i \in \mathrm{dom}\, F$, then there are a finite set C and a set $K \subseteq \{j : r(j) < r(i)\}$ critical in $(F \setminus A) \setminus C = F \setminus (A \cup C)$ such that $(F \setminus A)(i) \subseteq (F \setminus A)(K) \cup C = (F(K) \setminus A) \cup C$. Consequently $F(i) \subseteq F(K) \cup (A \cup C)$. Since $(A \cup C)$ is finite, we have shown that r is a rank function of F.

Families which possess a rank function will be characterized in the next section. In this section we generalize the results about Hall families to families which have a rank function.

Let $r: \mathrm{dom}\, F \to \gamma$ be a rank function of F. Obviously F has no marriage, if the following holds:

(1) There are an ordinal $\alpha < \gamma$ and a set $J \subset\subset I_{\alpha+1}$ such that $F \upharpoonright I_\alpha$ has a marriage and $F \upharpoonright I_\alpha \cup J$ has no marriage.

If $\gamma = 1$, then F is a Hall family, and (1) means that there is a $J \subset\subset \mathrm{dom}\, F$ such that $F \upharpoonright J$ has no marriage. Hence in this case the Compactness Theorem says that (1) holds if F has no marriage.

We want to generalize the Compactness Theorem by proving that (1) holds if F has no marriage. To do this, we associate Hall families with each rank function r of F.

Definition. Let H be a family such that $\mathrm{dom}\, F = \mathrm{dom}\, H$ and $H(i) \subset\subset F(i)$ for every $i \in \mathrm{dom}\, H$, and let $r: \mathrm{dom}\, H \to \mathrm{On}$ be a function. H is called a skeleton of F (with rank function r) if, for each $i \in \mathrm{dom}\, H$, there is a set $K \subseteq \{j : r(j) < r(i)\}$ critical in $F \setminus H(i)$ such that $F(i) \subseteq F(K) \cup H(i)$.

If F is a Hall family and $r: \mathrm{dom}\, F \to \gamma$ a function, then it is clear that F is a skeleton of F with rank function r.

Let r be a rank function of F. We want to show that there always exists a skeleton of F with rank function r.

Lemma 6.2. If there are a finite set A and a set K critical in $F \setminus A$ with $X \subseteq F(K) \cup A$, then there are sets $B \subset\subset X$ and $L \subseteq K$ critical in $F \setminus B$ such that $X \subseteq F(L) \cup B$.

Proof. Let $B \subseteq X \subseteq F(K) \cup A$ be a surplus with respect to $F \upharpoonright K$. By Fact 4, $|B| \leq |A|$. Therefore there is a finite maximal surplus B of X with respect to $F \upharpoonright K$. By Theorem 3.10, there is an $L \subseteq K$ critical in $F \setminus B$ such that $X \subseteq F(L) \cup B$.

Corollary 6.3. If r is a rank function of F, then there is a skeleton H of F with rank function r.

Let H be a skeleton of F with rank function r. We want to show that H has a marriage if $F \in \mathcal{M}$. Therefore we are going to investigate the connection between H and F.

Lemma 6.4.
Let $B \subset\subset X$, and let A be a maximal surplus of B with respect to F. Then the following conditions are equivalent:
(i) A is a maximal surplus of X with respect to F.
(ii) There are sets $C \subseteq B$ and K critical in $F \setminus C$ such that $X \subseteq F(K) \cup C$.

Proof. If A is a maximal surplus of X with respect to F, then, by Theorem 3.10, there is a set K critical in $F \setminus A$ such that $X \subseteq F(K) \cup A$. Take $C = A$ to get (ii). Now assume that (ii) holds. Then, by Theorem 3.10, there is a set L critical in $F \setminus A$ such that $B \subseteq F(L) \cup A$. By Fact 3, there is a set $N \subseteq K$ critical in $F \setminus (F(L) \cup A)$ such that $X \subseteq F(K) \cup C \subseteq F(N) \cup F(L) \cup A$. Since $N \cup L$ is critical in $F \setminus A$ by Fact 2, $X \setminus A \subseteq \mathrm{rng}\, f$ for every marriage f of $F \setminus A$. So we get (i), since A is a surplus of X with respect to F.

Let H be a skeleton of F with rank function r, let $r(i) = \alpha$, and suppose that $F \upharpoonright I_\alpha \cup \{i\}$ has a marriage f. By the definition of a skeleton and Lemma 6.4, there is a maximal surplus $A \subseteq H(i)$ of $F(i)$ with respect to $F \upharpoonright I_\alpha$, and Theorem 3.11 implies that $|A| \geq |\{f(i)\}|$. This means that we can find a marriage f' of $F \upharpoonright I_\alpha \cup \{i\}$ such that $f'(i) \in H(i)$. Since this holds for every α and i with $r(i) = \alpha$, we can hope that H has a marriage if F has a marriage. The next theorem will realize our hopes.

<u>Theorem 6.5</u>. Let H be a skeleton of F with rank function $r: \mathrm{dom}\, H \to \gamma$. If A is a finite surplus with respect to F, then A is a surplus with respect to H.

<u>Proof</u>. First by transfinite induction on α, $\alpha \leq \gamma$, and then by induction on $|L \setminus I_\alpha|$ we show that for every $L \subset\subset I_{\alpha+1}$ the following holds: If D is a finite surplus with respect to $F \upharpoonright I_\alpha \cup L$, then D is a surplus with respect to $H \upharpoonright I_\alpha \cup L$. Since $\mathrm{dom}\, F = \mathrm{dom}\, H = I_\gamma$, this will prove the theorem.

Let $L \subset\subset I_{\alpha+1}$ and D be finite such that $(F \upharpoonright I_\alpha \cup L) \setminus D$ has a marriage f. If $|L \setminus I_\alpha| = 0$, then we have to show that $H \upharpoonright I_\alpha \setminus D \in \mathcal{M}$. By the Compactness Theorem 1.4 it is sufficient to show that $H \upharpoonright N \setminus D \in \mathcal{M}$ for each $N \subset\subset I_\alpha$. Since N is finite and $I_\alpha = \{i : r(i) < \alpha\}$, there is a $\beta < \alpha$ with $N \subset\subset I_{\beta+1}$. By assumption, $(F \upharpoonright I_\beta \cup N) \setminus D$ has a marriage, and therefore the induction hypothesis implies that $(H \upharpoonright I_\beta \cup N) \setminus D \in \mathcal{M}$. Next assume that $|L \setminus I_\alpha| > 0$. If $i \in L \setminus I_\alpha$ and $N = L \setminus \{i\}$, then $D \cup \{f(i)\} \subseteq F(i) \cup D$ is a surplus with respect to $F \upharpoonright I_\alpha \cup N$. Let $E \supseteq D$ be a maximal surplus of $H(i) \cup D$ with respect to $F \upharpoonright I_\alpha \cup N$. By the definition of a skeleton, Fact 3 and Lemma 6.4, E is a maximal surplus of $F(i) \cup D$ with respect to $F \upharpoonright I_\alpha \cup N$. By Theorem 3.11, $|E| \geq |D \cup \{f(i)\}|$. Hence there exists an $a \in E \setminus D \subseteq H(i)$, and $D \cup \{a\}$ is a surplus with respect to $F \upharpoonright I_\alpha \cup N$. Therefore, by the induction hypothesis, it is a surplus with respect to $H \upharpoonright I_\alpha \cup N$. Let g be a marriage of $H \upharpoonright (I_\alpha \cup N) \setminus (D \cup \{a\})$. Then $g \cup \{(i,a)\}$ is a marriage of $(H \upharpoonright I_\alpha \cup L) \setminus D$, which completes the proof of the theorem.

Now we can prove the desired generalization of the Compactness Theorem 1.4.

Theorem 6.6. Let $r: \operatorname{dom} F \to \gamma$ be a rank function of F. If F has no marriage, then there are an ordinal $\alpha < \gamma$ and a set $L \subset\subset I_{\alpha+1}$ such that $F \upharpoonright I_\alpha$ has a marriage and $F \upharpoonright I_\alpha \cup L$ has no marriage.

Proof. Let H be a skeleton of F with rank function r. Since F has no marriage, H has no marriage. Hence there is a smallest $\beta \leq \gamma$ such that $H \upharpoonright I_\beta \notin \mathcal{M}$. By the Compactness Theorem 1.4, there is a set $L \subset\subset I_\beta$ such that $H \upharpoonright L \notin \mathcal{M}$. Now L is finite, and it follows that $L \subseteq I_{\alpha+1}$ for some $\alpha < \beta$. By the choice of β, $H \upharpoonright I_\alpha$ has a marriage, so $F \upharpoonright I_\alpha \in \mathcal{M}$. Since $H \upharpoonright I_\alpha \cup L$ has no marriage and $H \upharpoonright I_\alpha \cup L$ is a skeleton of $F \upharpoonright I_\alpha \cup L$, Theorem 6.5 implies that $F \upharpoonright I_\alpha \cup L$ has no marriage.

The generalization of Corollary 1.5 is:

Theorem 6.7. Let $r: \operatorname{dom} F \to \gamma$ be a rank function of F. If F has no marriage, then there are an ordinal $\alpha < \gamma$, a set $K \subseteq I_{\alpha+1}$ critical in F, and an $i \in I_{\alpha+1} \setminus (K \cup I_\alpha)$ such that $|K \setminus I_\alpha|$ is finite and $F(i) \subseteq F(K)$.

Proof. By Theorem 6.6, there are an $\alpha < \gamma$ and a set $L \subset\subset I_{\alpha+1}$ such that $F \upharpoonright I_\alpha \in \mathcal{M}$ and $F \upharpoonright I_\alpha \cup L \notin \mathcal{M}$. Since L is finite, there is a set $M \supseteq I_\alpha$ which is maximal matchable in $F \upharpoonright I_\alpha \cup L$. Then $M \neq I_\alpha \cup L$, and if we choose $i \in (I_\alpha \cup L) \setminus M$, Lemma 3.12 implies that there is a set $K \subseteq M \subseteq I_\alpha \cup L$ critical in F with $F(i) \subseteq F(K)$.

Let $r: \operatorname{dom} F \to \gamma$ be a rank function for the family F. If F has a marriage f, then, for every $\alpha < \gamma$ and every $J \subset\subset \{i : r(i) = \alpha\}$, $f[J] \subseteq F(J)$ is a surplus with respect to $F \upharpoonright I_\alpha$. Hence the following condition is necessary for the existence of a marriage of F:

(2) If $\alpha < \gamma$ such that $F \upharpoonright I_\alpha$ has a marriage and if $J \subset\subset I_{\alpha+1} \setminus I_\alpha$, then there is a surplus $B \subset\subset F(J)$ with respect to $F \upharpoonright I_\alpha$ such that $|B| \geq |J|$.

If $\gamma = 1$, then F is a Hall family and (2) says that $|F(J)| \geq |J|$ for every $J \subset\subset \operatorname{dom} F$. So in this case F has a marriage by Hall's Theorem. This shows that the following theorem is a generalization of Hall's Theorem.

Theorem 6.8. Let $r: \operatorname{dom} F \to \gamma$ be a rank function of a family F. F has a marriage if, for every $\alpha < \gamma$ with $F \upharpoonright I_\alpha \in \mathcal{M}$ and, for every $J \subset\subset I_{\alpha+1} \setminus I_\alpha$, there is a surplus $B \subseteq F(J)$ with respect to $F \upharpoonright I_\alpha$ such that $|B| \geq |J|$.

Proof. Let $\alpha < \gamma$ such that $F \upharpoonright I_\alpha \in \mathcal{M}$, and let $L \subset\subset I_{\alpha+1}$. By Theorem 6.6, it is sufficient to show that $F \upharpoonright I_\alpha \cup L$ has a marriage. Since L is finite, there exists a set $M \supseteq I_\alpha$ which is maximal matchable in $F \upharpoonright I_\alpha \cup L$. By Lemma 3.12, there is a set $K \subseteq M$ critical in F with $F((I_\alpha \cup L) \setminus M) \subseteq F(K)$. Let f be a marriage of $F \upharpoonright M$, and let $J = (K \setminus I_\alpha) \cup (L \setminus M)$. Then $f[K \setminus I_\alpha]$ is a maximal surplus of $F(J)$ with respect to $F \upharpoonright I_\alpha$. Theorem 3.11 and the assumption imply that $|f[K \setminus I_\alpha]| \geq |J| = |K \setminus I_\alpha| + |L \setminus M|$. Therefore $L \subseteq M$ and $F \upharpoonright I_\alpha \cup L \in \mathcal{M}$.

The following theorem is the generalization of Theorem 1.6 for families which possess a rank function.

Theorem 6.9. Let $r : \operatorname{dom} F \to \gamma$ be a rank function of the family F, and let B be a finite set. If $F \in \mathcal{M}$ and $B \subseteq \operatorname{rng} f$ for every marriage f of F, then there are an $\alpha < \gamma$ and a set $K \subseteq I_{\alpha+1}$ critical in F such that $B \subseteq F(K)$ and $|K \setminus I_\alpha| < \aleph_0$.

Proof. Let $b \in B$. By Lemma 6.1, r is a rank function of $F \setminus \{b\}$. Now $F \setminus \{b\}$ has no marriage, and Theorem 6.6 implies that there are an α_b and an $L_b \subset\subset I_{\alpha_b+1}$ such that $(F \upharpoonright I_{\alpha_b} \cup L_b) \setminus \{b\}$ has no marriage. Let $\alpha = \max\{\alpha_b : b \in B\}$, and let $L = \bigcup\{L_b : b \in B\}$. Since $F \upharpoonright I_\alpha \cup L \in \mathcal{M}$, the Main Lemma implies that there is a set $K \subseteq I_\alpha \cup L$ critical in F such that $F \upharpoonright (I_\alpha \cup L) \setminus \{a\}$ has a marriage for each $a \notin F(K)$. So we get $B \subseteq F(K)$ by the choice of α_b and L_b for $b \in B$.

The generalization of Theorem 1.7 is

Theorem 6.10. Let $r : \operatorname{dom} F \to \gamma$ be a rank function of F. If $B \subseteq X$ is a surplus with respect to F, then there is a maximal surplus $A \supseteq B$ of X with respect to F.

Proof. Let H be a skeleton of F with rank function r. The same proof as the one of Lemma 6.1 shows that $H \setminus Y$ is a skeleton of $F \setminus Y$ for every set Y. Hence, by Theorem 6.5, B is a surplus with respect to H. By Theorem 1.7, there is a maximal surplus $A \supseteq B$ of X with respect to H. Again by Theorem 6.5, A is a maximal surplus of X with respect to F.

Let $r : \mathrm{dom}\, F \to \gamma$ be a rank function of the family F. We cannot hope to find a maximal matchable set $M \supseteq J$ for every set $J \subseteq \mathrm{dom}\, F$ matchable in F. This is shown by the following example:

Let $\mathrm{dom}\, G = \omega+\omega$, $G(\alpha) = \{\alpha\}$ for $\alpha \in \omega$, and let $G(\alpha) = \omega$ for $\alpha \in \omega+\omega \setminus \omega$. Then $J = \omega+\omega\setminus\omega$ is matchable in G and $M \supseteq J$ is matchable in G if and only if $|\mathrm{dom}\, G \setminus M| = \aleph_0$. Therefore there is no maximal matchable $M \supseteq J$ in G. Clearly $M = \omega$ is maximal matchable.

We will show that there exists a maximal matchable set in F.

<u>Lemma 6.11</u>. If M is maximal matchable in F and K is critical in $F\setminus A$, then there is a set $L \subseteq M$ critical in $F \setminus A$ such that $F(K) \cup A \subseteq F(L) \cup A$.

<u>Proof</u>. By Lemma 3.12, there is a set $N \subseteq M$ critical in F with $F(\mathrm{dom}\, F \setminus M) \subseteq F(N)$, and Fact 3 implies that there is a set $N' \subseteq N \subseteq M$ critical in $F\setminus A$ such that $F(\mathrm{dom}\, F \setminus M) \subseteq F(N') \cup A$.

Again by Fact 3, there is a set $K' \subseteq K$ critical in $F\setminus(F(N') \cup A)$ such that $F(K) \cup A \subseteq F(N') \cup F(K') \cup A$. Since $F(\mathrm{dom}\, F \setminus M) \subseteq F(N') \cup A$, we have $K' \subseteq M$. If we take $L = N' \cup K'$, then $L \subseteq M$ and $F(K) \cup A \subseteq F(L) \cup A$. Finally, by Fact 2, L is critical in $F \setminus A$.

<u>Theorem 6.12</u>. If $r : \mathrm{dom}\, F \to \gamma$ is a rank function of F, then there is a set $M \subseteq \mathrm{dom}\, F$ such that $M \cap I_\alpha$ is maximal matchable in $F \restriction I_\alpha$ for each $\alpha \leq \gamma$.

<u>Proof</u>. Let H be a skeleton of F with rank function r. We choose, for each $\alpha \leq \gamma$, a maximal matchable set M_α in $H \restriction I_\alpha$ as follows:

Let $M_0 = \emptyset$. If α is a successor ordinal $\beta+1$, then, by Theorem 1.9, there is a maximal matchable $M_\alpha \supseteq M_\beta$ in $H \restriction I_\alpha$. If α is a limit ordinal, then let $M_\alpha = \bigcup\{M_\beta : \beta < \alpha\}$. We will show that $M = M_\gamma$ has the desired property. Since M_α is matchable in $H \restriction I_\alpha$, M_α is matchable in $F \restriction I_\alpha$. We prove by induction on α that M_α is maximal matchable in $F \restriction I_\alpha$.

Let $i \in I_\alpha$ such that $F \restriction M_\alpha \cup \{i\}$ has a marriage. By the induction hypothesis and Lemma 6.11, $H \restriction M_\alpha \cup \{i\}$ is a skeleton of $F \restriction M_\alpha \cup \{i\}$. So $H \restriction M_\alpha \cup \{i\}$ has a marriage by Theorem 6.5, and $i \in M_\alpha$ because M_α is maximal matchable in $H \restriction I_\alpha$. Therefore, M_α is maximal matchable in $F \restriction I_\alpha$.

This generalizes the last theorem of Section 1. In Section 2 we have defined the notion of a frame H of F. We want to show now that H is a frame of F if and only if H is a skeleton of F.

Theorem 6.13. If H is a skeleton of F with rank function $r:\operatorname{dom} H\to\gamma$, then H is a frame of F with rank function r.

Proof. Let $i\in L\subseteq\operatorname{dom} H$, $r(i)=\alpha$, and let $a\in F(i)$. We have to show that there is a $K\subset\subset I_\alpha$ such that $a\in H(L\cup K)$ and $|H(K)\setminus H(L)|\leq|K\setminus L|$. By Theorem 6.12, there is a set $M\subseteq\operatorname{dom} F$ such that $M_\beta := M\cap I_\beta$ is maximal matchable in $F\restriction I_\beta$ for every $\beta\leq\gamma$. Let A be a maximal surplus of $H(i)$ with respect to $F\restriction M_\alpha$. By Lemma 6.11 and the definition of a skeleton, there is a set $N\subseteq M_\alpha$ critical in $F\setminus H(i)$ with $F(i)\subseteq F(N)\cup H(i)$. Hence, by Lemma 6.4, A is a maximal surplus of $F(i)$ with respect to $F\restriction M_\alpha$. Again by Lemma 6.11, $H\restriction M_\alpha$ is a skeleton of $F\restriction M_\alpha$. So, by Theorem 6.5, A is a maximal surplus of $F(i)$ with respect to $H\restriction M_\alpha$. Take $L'=L\cap M_\alpha$. Then, by Lemma 2.1, there is a $K\subset\subset M_\alpha$ such that $a\in H(K\cup L')\cup H(i)\subseteq H(K\cup L)$ and $|H(K)\setminus H(L)|\leq$ $\leq|H(K)\setminus(H(L')\cup H(i))|\leq|K\setminus L'|=|K\setminus L|$.

To prove that every frame of F is a skeleton of F we need a technical lemma.

Lemma 6.14. Let H be a frame of F with rank function r. If M_α is maximal matchable in $H\restriction I_\alpha$ and $r(i)=\alpha$, then there is a maximal surplus $A\subseteq H(i)$ of $F(i)$ with respect to $H\restriction M_\alpha$.

Proof. Let A be a maximal surplus of $H(i)$ with respect to $H\restriction M_\alpha$, let $a\in F(i)\setminus A$, and let f be a marriage of $H\restriction M_\alpha\setminus A$. We want to show that $a\in\operatorname{rng} f$.

Let $N=\bigcup\{N'\subset\subset M_\alpha : |H(N')\setminus A|\leq|N'|\}$, and put $L=\{j : H(j)\subseteq H(N)\cup A\}$.

Claim. $I_\alpha\setminus L\subseteq M_\alpha\setminus N$.

Proof. $N\subseteq L$ by the definition of L. Let $j\in I_\alpha\setminus M_\alpha$, and let h be a marriage of $H\restriction M_\alpha\setminus A$. Since M_α is maximal matchable in $H\restriction I_\alpha$, $H(j)\setminus A\subseteq\operatorname{rng} h$. Hence $j\in L$ by Theorem 1.6, which proves the claim.

Since A is a maximal surplus of $H(i)$ with respect to $H\restriction M_\alpha$, we have $i\in L$ by Theorem 1.6 again. By the definition of a frame, there is a set $K \subset\subset I_\alpha$ such that $a \in H(K \cup L) \setminus A$ and $|H(K) \setminus H(L)| \leq |K \setminus L|$. Now $K \setminus L \subseteq I_\alpha \setminus L \subseteq M_\alpha \setminus N$ and $f[N] = H(N) \setminus A = H(L) \setminus A$, so we get:

$$|K \setminus L| = |f[K \setminus L]| \leq |(H(K) \setminus A) \setminus (H(L) \setminus A)| \leq |H(K) \setminus H(L)| \leq |K \setminus L|.$$

Hence $f[K \cup L] = H(L \cup K) \setminus A$, and $a \in \operatorname{rng} f$.

Theorem 6.15. If H is a frame of F with rank function $r : \operatorname{dom} H \to \gamma$, then H is a skeleton of F with rank function r.

Proof. Let $r(i) = \alpha$. We have to find a set K critical in $F \restriction I_\alpha \setminus H(i)$ such that $F(i) \subseteq F(K) \cup H(i)$.

As in Theorem 6.12, we choose a continuous chain $(M_\alpha : \alpha \leq \gamma)$ such that M_α is maximal matchable in $F \restriction I_\alpha$ for every $\alpha \leq \gamma$. By Lemma 6.14 and Lemma 2.1, $H \restriction M_\alpha$ is a frame of $F \restriction M_\alpha$. Again by Lemma 6.14, there is a maximal surplus $A \subseteq H(i)$ of $F(i)$ with respect to $H \restriction M_\alpha$. Since $H \restriction M_\alpha$ is a frame of $F \restriction M_\alpha$, Theorem 2.3 shows that A is a maximal surplus of $F(i)$ with respect to $F \restriction M_\alpha$. By Theorem 3.10, there is a set $L \subseteq M_\alpha$ critical in $F \setminus A$ with $F(i) \subseteq F(L) \cup A$. Now Fact 3 guarantees the existence of a set $K \subseteq L$ critical in $F \setminus H(i)$ such that $F(i) \subseteq F(L) \cup A \subseteq F(K) \cup H(i)$.

Corollary 6.16.
The following conditions are equivalent:
(1) F has a rank function.
(2) F has a frame.

Proof. If r is a rank function of F, then, by Corollary 6.3, there is a skeleton H of F with rank function r. By Theorem 6.13, H is a frame of F. Conversely, if H is a frame of F with rank function r, then, by Theorem 6.15, H is a skeleton of F with rank function r. By the definition of a skeleton, r is a rank function of F.

§7. Stable families

A family F is called stable if, for each $i \in I$, there are a finite set A and a set K critical in $F \setminus A$ such that $F(i) \subseteq A \cup F(K)$.

Clearly every Hall family and every critical family is stable. One could say that a stable family is a Hall family "modulo critical sets". We want to prove that the following conditions are equivalent:

(a) F is stable.

(b) F has a rank function.

By Corollary 6.16, (b) is equivalent to the assertion "F has a frame".

Thus we can construct stable families from Hall families H as follows. Take a function $r : \mathrm{dom}\, H \to \gamma$. Then extend $H(i)$ to $F(i)$ by joining women a with the following property:

For any $L \subseteq \mathrm{dom}\, H$ with $i \in L$, there is a set $K \subset\subset \{j : r(j) < r(i)\}$ such that $a \in H(K \cup L)$ and $|H(K) \setminus H(L)| \leq |K \setminus L|$.

Roughly speaking, the circle of girl friends $F(i)$ of the man i may become very large, but when he wants to marry, most of his girl friends are already promised to a man of lower rank, so that he can restrict his attention to $H(i)$ if he wishes that every man of the family F is getting married.

(b) implies (a) by definition. So we have to prove that (a) implies (b). Let F be a family. By recursion, we define sets $I_\alpha \subseteq \mathrm{dom}\, F$ as follows:

$$I_\alpha = \{i : \text{There are a } \beta < \alpha \text{, a finite set } A \text{, and a set } K \subseteq I_\beta \text{ critical in } F \setminus A \text{ such that } F(i) \subseteq F(K) \cup A\}$$

By the definition of I_α we have:

1. $I_0 = \emptyset$.
2. $I_1 = \{i : F(i) \text{ is finite}\}$.
3. $I_\beta \subseteq I_\alpha$ for $\beta \leq \alpha$.
4. $I_\alpha = \bigcup\{I_\beta : \beta < \alpha\}$ for every limit ordinal α.
5. There is an α with $I_{\alpha+1} = I_\alpha$
6. If $I_{\alpha+1} = I_\alpha$, then $I_\beta = I_\alpha$ for every $\beta \geq \alpha$.

<u>Let γ be the smallest ordinal α with $I_{\alpha+1} = I_\alpha$.</u>

By the definition of I_β, $\beta \leq \gamma$, there is a rank function r of $F \upharpoonright I_\gamma$ with $r(i) = \beta$ if and only if $i \in I_{\beta+1} \setminus I_\beta$. This rank function is called the <u>canonical rank function of $F \upharpoonright I_\gamma$</u>.

<u>Lemma 7.1</u>. Let r be the canonical rank function of $F \upharpoonright I_\gamma$ and s be an arbitrary rank function of $F \upharpoonright \mathrm{dom}\, s$. Then $\mathrm{dom}\, s \subseteq I_\gamma$ and $r(i) \leq s(i)$ for each $i \in \mathrm{dom}\, s$.

<u>Proof</u>. By the definition of r, $I_\alpha = \{i : r(i) < \alpha\}$. Hence it is sufficient to prove by induction on α that $\{i : s(i) < \alpha\} \subseteq I_\alpha$ for each ordinal α. Let $s(i) = \beta < \alpha$. Then, by the definition of a rank function and by the induction hypothesis, there are a finite set A and a set $K \subseteq \{j : s(j) < \beta\} \subseteq I_\beta$ critical in $F \setminus A$ such that $F(i) \subseteq F(K) \cup A$. This implies $i \in I_{\beta+1} \subseteq I_\alpha$.

Let F be a stable family. If we can show that $I_\gamma = \mathrm{dom}\, F$, then the canonical rank function in F is a rank function of F. This will prove that (a) implies (b). We begin our considerations with the following theorem.

<u>Theorem 7.2</u>. Let A be finite, and let J be countable. If $F \upharpoonright I_\gamma \setminus A$ and $F \setminus J$ have a marriage, then $F \setminus A \in \mathcal{M}$.

<u>Proof</u>. In this proof we call a finite matching g in F admissible if $F \upharpoonright I_\gamma \setminus A \setminus g$ has a marriage. $\emptyset$ is admissible, since $F \upharpoonright I_\gamma \setminus A \in \mathcal{M}$. As usual we prove the following

<u>Claim</u>. If g is admissible and $K \subset\subset \mathrm{dom}\, F$, then there is an admissible $h \supseteq g$ such that $K \subseteq \mathrm{dom}\, h$.

We prove the claim by induction on $|K \setminus \mathrm{dom}\, g|$. If $|K \setminus \mathrm{dom}\, g| = 0$, then let $h = g$. Otherwise choose $i \in K \setminus \mathrm{dom}\, g$. Then, by the induction hypothesis, there is an admissible $h_1 \supseteq g$ with $K \setminus \{i\} \subseteq \mathrm{dom}\, h_1$. If $i \in \mathrm{dom}\, h_1$, then let $h = h_1$. So assume that $i \notin \mathrm{dom}\, h_1$. By the Main Lemma (Theorem 3.4), there is a set $L \subseteq I_\gamma$ critical in $F \setminus (A \cup \mathrm{rng}\, h_1)$ such that $F \upharpoonright I_\gamma \setminus (A \cup \{a\}) \setminus h_1$ has a marriage for each $a \notin F(L) \setminus (A \cup \mathrm{rng}\, h_1)$. If $F(i) \subseteq F(L) \cup (A \cup \mathrm{rng}\, h_1)$, then $i \in I_{\gamma+1} = I_\gamma$. In this case let f be a marriage of $F \upharpoonright I_\gamma \setminus A \setminus h_1$ and take $h = h_1 \cup \{(i, f(i))\}$. Otherwise let $a \in F(i) \setminus (F(L) \cup A \cup \mathrm{rng}\, h_1)$ and $h = h_1 \cup \{(i, a)\}$. This proves the claim.

Now we can finish the proof as follows. Choose $J_n \subset\subset J$ for each $n \in \omega$ such that $\bigcup\{J_n : n \in \omega\} = J$, and let f be a marriage of $F \setminus J$. Put $g_0 = \emptyset$ and suppose that we have already chosen an admissible matching g_n. Take $K_n = \{i : f(i) \in \mathrm{rng}\, g_n \cup A\} \cup J_n$. Then the claim yields an admissible matching $g_{n+1} \supseteq g_n$ such that $K_n \subseteq \mathrm{dom}\, g_{n+1}$. If we take $g = \bigcup\{g_n : n \in \omega\}$, then $f \upharpoonright (\mathrm{dom}\, f \setminus \mathrm{dom}\, g) \cup g$ is a marriage of $F \setminus A$.

<u>Lemma 7.3</u>. If there is a finite set A such that $F \setminus A$ is critical, then F has a rank function.

Proof. Let r be the canonical rank function of $F \upharpoonright I_\gamma$. It is sufficient to show that $\operatorname{dom} F = I_\gamma$. By the Main Lemma (Theorem 3.4), there is a set $L \subseteq I_\gamma$ critical in $F \setminus A$ such that $F \upharpoonright I_\gamma \setminus (A \cup \{a\})$ has a marriage for each $a \notin F(L) \setminus A$. Since $F \setminus A \in \mathcal{M}$, Theorem 7.2 implies that $F \setminus (A \cup \{a\})$ has a marriage for every $a \notin F(L) \setminus A$. Now $F \setminus A$ is critical, so $F(\operatorname{dom} F) \subseteq F(L) \cup A$. It follows that $\operatorname{dom} F \subseteq I_{\gamma+1} = I_\gamma$.

Definition. Let F be a family. We define $\underline{I_F} = \{i : \text{There are a finite set } A \text{ and a set } K \text{ critical in } F \setminus A \text{ such that } F(i) \subseteq F(K) \cup A\}$.

Theorem 7.4. $I_F = I_\gamma$ for every family F.

Proof. $I_\gamma \subseteq I_F$ by the definitions of I_γ and I_F. Let $i \in I_F$. Then there are a finite set A and a set K critical in $F \setminus A$ such that $F(i) \subseteq F(K) \cup A$. By Lemma 7.3, $F \upharpoonright K$ has a rank function s, and by Lemma 7.1, $K = \operatorname{dom} s \subseteq I_\gamma$. Therefore $i \in I_{\gamma+1} = I_\gamma$.

Corollary 7.5.
The following conditions are equivalent:
(1) $F \upharpoonright I_F$ has a marriage.
(2) If K is critical in F and $F(i) \subseteq F(K)$, then $i \in K$.

Proof. If K is critical in F and $F(i) \subseteq F(K)$, then $K \cup \{i\} \subseteq I_F$. But $F \upharpoonright I_F$ has a marriage, so $i \in K$. For the proof of the converse, let r be the canonical rank function of $F \upharpoonright I_\gamma = F \upharpoonright I_F$. (2) and Theorem 6.7 imply that $F \upharpoonright I_F$ has a marriage.

Corollary 7.6.
If $\operatorname{dom} F \setminus I_F$ is countable, then the following conditions are equivalent:
(a) F has a marriage.
(b) $F \upharpoonright I_F$ has a marriage.

Proof. It is obvious that (a) implies (b). Let $F \upharpoonright I_F \in \mathcal{M}$. Since $I_F = I_\gamma$ by Theorem 7.4, $\operatorname{dom} F \setminus I_\gamma$ is countable. Therefore F has a marriage by Theorem 7.2.

Now we turn back to stable families. If F is stable, then dom F = $= I_F = I_\gamma$, hence the canonical rank function of $F \upharpoonright I_\gamma$ is a rank function of F. We call this rank function the canonical rank function of F. Therefore the following conditions are equivalent:

(3)
(a) F is stable.
(b) F has a rank function.

Now we want to reformulate the results of Section 6 for stable families. Let F be stable, and let r be a rank function of F. We define $r(\text{dom}\,F) = \sup\{r(i)+1 : i \in \text{dom}\,F\}$ and $rk(F) = \min\{r(\text{dom}\,F) : r$ is a rank function of F$\}$. Then the following holds:

1. $rk(F) = 0 \leftrightarrow F = \emptyset$.
2. $rk(F) = 1 \leftrightarrow F$ is a Hall family.

Further Lemma 7.1 implies

3. If r is the canonical rank function of F, then $r(\text{dom}\,F) = rk(F)$.

A generalization of Hall's Theorem can be formulated as follows:

Corollary 7.7. A stable family F has a marriage if, for each $J \subseteq \text{dom}\,F$ with $rk(F \upharpoonright J) < rk(F)$ and every $L \subset\subset \text{dom}\,F \setminus J$, there is a surplus $B \subseteq F(L)$ with respect to $F \upharpoonright J$ such that $|B| \geq |L|$.

Proof. By assumption, F has a rank function $r : \text{dom}\,F \to rk(F)$. We want to apply Theorem 6.8 to show that F has a marriage. Let $\alpha < rk(F)$ such that $F \upharpoonright \{i : r(i) < \alpha\}$ has a marriage, and let $L \subset\subset \{i : r(i) = \alpha\}$. We have to find a surplus $B \subseteq F(L)$ with respect to $F \upharpoonright \{i : r(i) < \alpha\}$ such that $|B| \geq |L|$. If $J = \{i : r(i) < \alpha\}$, then $r \upharpoonright J$ is a rank function of $F \upharpoonright J$ and $rk(F \upharpoonright J) \leq \alpha < rk(F)$. Therefore there is a surplus $B \subseteq F(L)$ with respect to $F \upharpoonright J$ with $|B| \geq |L|$. Now $F \in \mathcal{M}$ by Theorem 6.8.

Definition. Let F be a stable family, let r be the canonical rank function of F, and let $J \subseteq \text{dom}\,F$. We write $J \sqsubset\sqsubset \text{dom}\,F$, if there is a set $L \subset\subset J$ such that $rk(F \upharpoonright (J \setminus L)) < rk(F)$. Note that if $rk(F) = 1$, then $J \sqsubset\sqsubset \text{dom}\,F$ if and only if $J \subset\subset \text{dom}\,F$.

Just like Corollary 7.7 (using Theorem 6.6), we can prove the following generalization of the Compactness Theorem:

Corollary 7.8. A stable family F has no marriage if there is a set $J \sqsubset\sqsubset \mathrm{dom}\, F$ such that $F \restriction J$ has no marriage.

If we apply Theorem 6.7, we get the following generalization of Corollary 1.5.

Corollary 7.9. If a stable family F has no marriage, then there are a set $K \sqsubset\sqsubset \mathrm{dom}\, F$ critical in F and an $i \in \mathrm{dom}\, F \setminus K$ such that $F(i) \subseteq F(K)$.

Theorem 6.9 can be reformulated as follows:

Corollary 7.10. Let $F \in \mathcal{M}$ be a stable family, and let B be a finite set. If $B \subseteq \mathrm{rng}\, f$ for every marriage f of F, then there is a set $K \sqsubset\sqsubset \mathrm{dom}\, F$ critical in F such that $B \subseteq F(K)$.

This is a generalization of Theorem 1.6. From Theorem 6.10 we get the following generalization of Theorem 1.7.

Corollary 7.11. If F is a stable family and $B \subseteq X$ is a surplus with respect to F, then there is a maximal surplus $A \supseteq B$ of X with respect to F.

To generalize the last Theorem of Section 1 we call a set $J \subseteq \mathrm{dom}\, F$ stable in F if $F \restriction J$ is a stable family. If F is a Hall family, then every $J \subseteq \mathrm{dom}\, F$ is stable in F.

Corollary 7.12. Let F be stable, and let $J \subseteq \mathrm{dom}\, F$ be stable and matchable in F. Then there is a set $M \supseteq J$ which is stable and maximal matchable in F.

Proof. By Theorem 7.4, there are rank functions $s : J \to \beta$ of $F \restriction J$ and $t : \mathrm{dom}\, F \to \delta$ of F. We define $r : \mathrm{dom}\, F \to \gamma$, where $\gamma = \beta + \delta$, by

$$r(i) = \begin{cases} s(i) & \text{if } i \in J, \\ \beta + t(i) & \text{if } i \in \operatorname{dom} F \setminus J. \end{cases}$$

Then r is a rank function of F. By Theorem 6.12, there is a set $M \subseteq \operatorname{dom} F$ such that $M_\alpha = M \cap \{i : r(i) < \alpha\}$ is maximal matchable in $F \restriction \{i : r(i) < \alpha\}$ for every $\alpha \leq \gamma$. Since $\{i : r(i) < \gamma\} = \operatorname{dom} F$, $M = M_\gamma$ is maximal matchable in F. Since $J = \{i : r(i) < \beta\}$ is matchable in F, we get that $J = M_\beta \subseteq M$. By Lemma 6.11, $M = M_\gamma$ is stable in F.

In Section 4 we had defined a hierarchy $\mathcal{F}_\alpha$ of families such that $F \in \mathcal{F}_\alpha$ for some α if and only if F is critical. We want to show that this hierarchy is a natural one. For this reason we show:

$\mathcal{F}_\alpha = \{F : F \text{ is critical and } rk(F) \leq \alpha\}$.

To do this we notice first that for each finite set A

(4) $rk(F) = rk(F \setminus A)$.

This follows from Lemma 6.1.

(5) If $\mathcal{K}$ is a set of stable sets in F such that $\bigcup\mathcal{K} = \operatorname{dom} F$, then $rk(F) \leq \sup\{rk(F \restriction K) : K \in \mathcal{K}\}$.

Proof. By assumption, there is a rank function $r_K : K \to rk(K)$ for every $K \in \mathcal{K}$. Let r be the canonical rank function of $F \restriction I_\gamma$. Then (5) follows from Lemma 7.1.

From (5) we get for stable families F:

(6) If $L, K \sqsubset\sqsubset \operatorname{dom} F$, then $L \cup K \sqsubset\sqsubset \operatorname{dom} F$.

Theorem 7.13. $\mathcal{F}_\alpha = \{F : F \text{ is critical and } rk(F) \leq \alpha\}$

Proof. We prove the theorem by induction on α. First let $F \in \mathcal{F}_\alpha$. Then F is critical and there is a directed system $\mathcal{K}$ such that $\bigcup\mathcal{K} = \operatorname{dom} F$ and $F \restriction K$ is α-finite for each $K \in \mathcal{K}$. By (5), it is sufficient to show that $rk(F \restriction K) \leq \alpha$. Let $K \in \mathcal{K}$. Then there is a set $L \subset\subset K$ and an $A \subset\subset F(L)$ such that $F(L) \subseteq F(K \setminus L) \cup A$ and $F \restriction (K \setminus L) \setminus A \in \mathcal{F}_\beta$ for some $\beta < \alpha$. By the induction hypothesis, $rk(F \restriction (K \setminus L) \setminus A) = \beta$, and by (4), $rk(F \restriction (K \setminus L)) = \beta < \alpha$. So there is a rank function $r : (K \setminus L) \to \beta$ of $F \restriction (K \setminus L)$. Now $F \restriction (K \setminus L) \setminus A$ is critical and $F(L) \subseteq F(K \setminus L) \cup A$. Therefore $s = r \cup \{(i, \beta) : i \in L\}$ is a rank function of $F \restriction K$ and

$rk(F \restriction K) \leq \beta + 1 \leq \alpha$.

For the proof of the converse inclusion let F be critical with $rk(F) \leq \alpha$, and let $\mathcal{K} = \{K : K \sqsubset\sqsubset \operatorname{dom} F$ is critical in $F\}$. If $K, L \in \mathcal{K}$, then $K \cup L \sqsubset\sqsubset \operatorname{dom} F$ by (6). Since F has a marriage, $K \cup L$ is critical. Hence $\mathcal{K}$ is a directed system.

Claim 1. $\bigcup \mathcal{K} = \operatorname{dom} F$.

Proof. Let $a \in F(\operatorname{dom} F)$. Then, by Corollary 7.10, there is a set $K \sqsubset\sqsubset \operatorname{dom} F$ critical in F with $a \in F(K)$. This implies $F(\bigcup \mathcal{K}) = F(\operatorname{dom} F)$. If we choose a marriage f of F, then $f[\bigcup \mathcal{K}] = F(\bigcup \mathcal{K}) = F(\operatorname{dom} F) = \operatorname{rng} f$, and so $\bigcup \mathcal{K} = \operatorname{dom} F$.

Claim 2. $F \restriction K$ is α-finite for any $K \in \mathcal{K}$.

Proof. Let $L \subset\subset K$ such that $rk(F \restriction (K \setminus L)) = \beta < \alpha$, let f be a marriage of $F \restriction K$, and let $A = f[L]$. Since $F \restriction K$ is critical and $F(L) \subseteq F(K \setminus L) \cup A$, $F \restriction (K \setminus L) \setminus A$ is critical and $(F(i) \cap A : i \in L)$ is critical. Further $rk(F \restriction (K \setminus L) \setminus A) = rk(F \restriction K \setminus L) = \beta$. Hence, by the induction hypothesis, $(F \restriction (K \setminus L) \setminus A) \in \mathcal{F}_\beta$. So $F \restriction K$ is α-finite.

The two claims imply that $F \in \mathcal{F}_\alpha$.

§8. Margin functions and demand functions

In this section we present variations of the criteria of [N2], [N3], and [HPS2].

In the last section we had defined sets $I_\alpha \subseteq \operatorname{dom} F$ by recursion as follows:

$$I_\alpha = \{i : \text{There are a } \beta < \alpha, \text{ a finite set } A, \text{ and a set } K \subseteq I_\beta \text{ critical in } F \setminus A \text{ such that } F(i) \subseteq F(K) \cup A\}.$$

By Lemma 6.2 we have

$$I_\alpha = \{i : \text{There are a } \beta < \alpha, \text{ an } A \subset\subset F(i), \text{ and a set } K \subseteq I_\beta \text{ critical in } F \setminus A \text{ such that } F(i) \subseteq F(K) \cup A\}.$$

Let X be a set of women. We want to calculate the maximal number of women in X whom we can leave unmarried after we have married each man in I_α.

Therefore we define so-called <u>margin functions</u> m_α for arbitrary sets X as follows:

$m_\alpha(X) = \sup\{|B| + 1 : B \subset\subset X$ is a surplus with respect to $F \restriction I_\alpha\} - 1$, where we put $\omega - 1 = \omega$.

If $m_\alpha(X)\geq 0$, then $\emptyset$ is a surplus with respect to $F\upharpoonright I_\alpha$, and therefore $F\upharpoonright I_\alpha$ has a marriage. Otherwise $F\upharpoonright I_\alpha$ has no marriage.

Let $m_\alpha(X)\in\omega$. Then there is a surplus $B\subseteq X$ with respect to $F\upharpoonright I_\alpha$ such that $|B| = m_\alpha(X)$, and so B must be a maximal surplus of X with respect to $F\upharpoonright I_\alpha$. Conversely, if C is a finite maximal surplus of X with respect to $F\upharpoonright I_\alpha$, then $m_\alpha(X) = |C|$ by Theorem 3.11.

If $m_\alpha(X) = \omega$ and $B\subset\subset X$ is a surplus with respect to $F\upharpoonright I_\alpha$, then, again by Theorem 3.11, there is an $a\in X\setminus B$ such that $B\cup\{a\}$ is a surplus of X with respect to $F\upharpoonright I_\alpha$.

With help of the functions m_α we can reformulate the results of the last section.

<u>Corollary 8.1</u>.
The following conditions are equivalent:
(1) $F\upharpoonright I_F$ has a marriage.
(2) $m_\alpha(\emptyset)\geq 0$ for every ordinal α.

<u>Proof</u>. By Theorem 7.5, we have $I_F = I_\gamma$. So $F\upharpoonright I_F\in\mathcal{M}$ implies $F\upharpoonright I_\alpha\in\mathcal{M}$, and consequently $m_\alpha(\emptyset)\geq 0$ for every ordinal α. To prove the converse, we observe that (2) implies especially $m_\gamma(\emptyset)\geq 0$. Therefore $F\upharpoonright I_\gamma = F\upharpoonright I_F$ has a marriage.

<u>Corollary 8.2</u>.
If F is a countable family, then the following conditions are equivalent:
(1) F has a marriage.
(2) $m_\alpha(\emptyset)\geq 0$ for every ordinal α.

<u>Proof</u>. Since F is countable, $\operatorname{dom} F\setminus I_F$ is countable. So the corollary follows from Corollary 7.6 and Corollary 8.1.

<u>Corollary 8.3</u>.
The following conditions are equivalent:
(1) F is critical.
(2) There is an ordinal β such that $m_\alpha(F(\operatorname{dom} F)) = 0$ for every $\alpha\geq\beta$.

Proof. If F is critical, then $I_F = \operatorname{dom} F$. Hence $I_F = I_\gamma$ by Theorem 7.4, and consequently $\operatorname{dom} F = I_\alpha$ for each $\alpha \geq \gamma$. Further we have $m_\alpha(F(\operatorname{dom} F)) = 0$ for every $\alpha \geq \gamma$, since F is critical. Now assume (2) and let $\alpha > \gamma$ with $m_\alpha(F(\operatorname{dom} F)) = 0$. Then $I_\gamma = I_\alpha$, and therefore $m_\gamma(F(\operatorname{dom} F)) = 0$ and $F \upharpoonright I_\gamma$ has a marriage. Since $0 = m_\gamma(F(\operatorname{dom} F)) \geq m_\gamma(F(I_\gamma))$, $F \upharpoonright I_\gamma$ is critical and $F(\operatorname{dom} F) \subseteq F(I_\gamma)$. Hence $\operatorname{dom} F = I_{\gamma+1} = I_\alpha$, which implies that F is critical.

Next we want to show that we can calculate the values of $m_\alpha(X)$. First consider m_0. Since $I_0 = \emptyset$, it follows that

$$m_0(X) = \begin{cases} |X| & \text{if } X \text{ is finite,} \\ \omega & \text{otherwise} \end{cases}.$$

Now we calculate $m_1(X)$. If Z is finite, let

$$E_0(Z) = \{i : F(i) \subseteq Z\}$$

Claim 1. $E_0(Z) \subseteq I_1$.

Proof. If $F(i) \subseteq Z$, then $F(i)$ is finite and so $i \in I_1$.

Definition. We say that B can be omitted (in F) at stage 1, if, for every finite $Z \supseteq B$,

$$|E_0(Z)| + |B| \leq |Z|.$$

Claim 2.

The following conditions are equivalent:

(1) B is a surplus with respect to $F \upharpoonright I_1$.

(2) B can be omitted at stage 1.

Proof. Assume (1), let f be a marriage of $F \upharpoonright I_1 \setminus B$, and let $Z \supseteq B$ be finite. Then $B \cup f[E_0(Z)] \subseteq Z$ and $E_0(Z) \subseteq I_1$. Hence $|B| + |E_0(Z)| \leq |Z|$.

To prove the converse direction, we want to apply Hall's Theorem. For each $L \subset\subset I_1$, we have to show that $|F(L) \setminus B| \geq |L|$. Let $Z = F(L) \cup B$. Since B can be omitted at stage 1, we get

$$|L| + |B| \leq |E_0(Z)| + |B| \leq |Z| = |F(L) \setminus B| + |B|.$$

So $|L| \leq |F(L) \setminus B|$, and the claim is proved.

From this claim it follows that

$$m_1(X) = \sup\{|B| + 1 : B \subset\subset X \text{ can be omitted at stage } 1\} - 1.$$

Now we turn our attention to the general case.

Lemma 8.4. If $F \restriction I_\alpha$ has a marriage, then $I_\beta = \{i : \text{There is a } \delta<\beta$ such that $m_\delta(F(i)) \in \omega\}$ for each $\beta \leq \alpha+1$.

Proof. Let $i \in I_\beta$. Then there are a $\delta<\beta$, an $A \subset\subset F(i)$, and a set $K \subseteq I_\delta$ critical in $F \setminus A$ such that $F(i) \subseteq F(K) \cup A$. Since $F \restriction I_\delta$ has a marriage, there is a maximal surplus $C \subseteq A$ with respect to $F \restriction I_\delta$. By Lemma 6.4, C is a maximal surplus of F(i) with respect to $F \restriction I_\delta$, and consequently $m_\beta(F(i)) = |C| \in \omega$. To prove the converse inclusion, let $\delta<\beta$ and $m_\delta(F(i)) \in \omega$. Then there is a maximal surplus $C \subset\subset F(i)$ with respect to $F \restriction I_\delta$, and $m_\delta(F(i)) = |C|$. By Theorem 3.10, there is a set $K \subseteq I_\delta$ critical in $F \restriction I_\delta \setminus C$ with $F(i) \subseteq F(K) \cup C$, which implies that $i \in I_\beta$.

Definition. If $m_\alpha(Z) \in \omega$, we define

$$E_\alpha(Z) = \{i \notin I_\alpha : F(i) \subseteq Z\}.$$

Similar to Claim 1 we show:

Lemma 8.5. If $m_\alpha(Z) \in \omega$, then $E_\alpha(Z) \subseteq I_{\alpha+1}$.

Proof. Since $m_\alpha(Z) \geq 0$, $F \restriction I_\alpha \in \mathcal{M}$. Let $i \in E_\alpha(Z)$, and let $B \subseteq F(i) \subseteq Z$ be a surplus with respect to $F \restriction I_\alpha$. Then $|B| \leq m_\alpha(Z)$, and so there is a finite maximal surplus C of F(i) with respect to $F \restriction I_\alpha$. By definition, $m_\alpha(F(i)) = |C| \in \omega$, which implies that $i \in I_{\alpha+1}$ by Lemma 8.4.

Definition. We say that a finite set B can be <u>omitted at stage α</u>, if, for any $\beta<\alpha$ and any $Z \supseteq B$ with $m_\beta(Z) \in \omega$,

$$|E_\beta(Z)| + |B| \leq m_\beta(Z).$$

Similar to Claim 2 we show:

Theorem 8.6.
For each ordinal α and for each finite set B, the following conditions are equivalent:

(1) B is a surplus with respect to $F \restriction I_\alpha$.
(2) B can be omitted at stage α. |

Proof. Assume (1), let $\beta<\alpha$ and $Z \supseteq B$ with $m_\beta(Z)\in\omega$, and let f be a marriage of $F \restriction I_\alpha \setminus B$. Then $f[E_\beta(Z)] \cup B$ is a surplus of Z with respect to $F \restriction I_\beta$. By Lemma 8.5, we get $E_\beta(Z) \subseteq \operatorname{dom} f$, so

$$|E_\beta(Z)| + |B| \leq m_\beta(Z).$$

The proof of the converse implication is more complicated. Let r be the canonical rank function of $F \restriction I_\alpha$. Then $\{i : r(i) < \beta\} = I_\beta$ for each $\beta<\alpha$. By Lemma 6.1, r is a rank function of $F \restriction I_\alpha \setminus B$. We want to apply Theorem 6.8 to show that $F \restriction I_\alpha \setminus B$ has a marriage.

Let $\beta<\alpha$ such that $F \restriction I_\beta \setminus B \in \mathcal{M}$, and let $L \subset\subset I_{\beta+1} \setminus I_\beta$. We have to show that there is a set $A \subseteq F(L) \setminus B$ such that $|A| \geq |L|$ and $F \restriction I_\beta \setminus (A \cup B) \in \mathcal{M}$. Take $Z = F(L) \cup B$ and $i \in L$. Then $F(i)$ has a finite maximal surplus with respect to $F \restriction I_\beta$. Therefore, by the pigeonhole principle, Z has a finite maximal surplus with respect to $F \restriction I_\beta$, and $m_\beta(Z) \in \omega$. If A is a maximal surplus of $Z \setminus B$ with respect to $F \restriction I_\beta \setminus B$, then $A \cup B$ is a maximal surplus of Z with respect to $F \restriction I_\beta$, and so $m_\beta(Z) = |A| + |B|$. Since $L \subseteq E_\beta(Z)$, we get

$$|L| + |B| \leq |E_\beta(Z)| + |B| \leq m_\beta(Z) = |A| + |B|.$$

From this follows $|L| \leq |A|$, since $|B|$ is finite. Now Theorem 6.8 implies that $F \restriction I_\alpha \setminus B \in \mathcal{M}$, and the theorem is proved.

Definition. We define functions l_α by recursion as follows. First let $J_\alpha := \{i : \text{There is a } \beta<\alpha \text{ with } l_\beta(F(i)) \in \omega\}$. We say that B can be left unmarried at stage α if $|\{i \notin J_\beta : F(i) \subseteq Z\}| + |B| \leq l_\beta(Z)$ for any $\beta<\alpha$ and any $Z \supseteq B$ with $l_\beta(Z) \in \omega$. Then

$$l_\alpha(X) := \sup\{|B| + 1 : B \subset\subset X \text{ can be left unmarried at stage } \alpha\} - 1.$$

Theorem 8.7. $l_\alpha(X) = m_\alpha(X)$ for every ordinal α and every finite set X.

Proof. We prove the theorem by induction on α. A set $B \subset\subset X$ can be left unmarried at stage α if $|\{i \notin J_\beta : F(i) \subseteq Z\}| + |B| \leq l_\beta(Z)$ for every $\beta < \alpha$ and every $Z \supseteq B$ with $l_\beta(Z) \in \omega$. By the induction hypothesis, $l_\beta(Z) = m_\beta(Z)$ for each $\beta < \alpha$. Therefore $F \upharpoonright I_\beta$ has a marriage, and, by Lemma 8.4, $I_\beta = \{i : \text{There is a } \delta < \beta \text{ with } m_\delta(F(i)) \in \omega\}$. By the induction hypothesis again, $m_\delta(F(i)) = l_\delta(F(i))$ for $\delta < \beta < \alpha$. Hence $I_\beta = J_\beta$. Consequently $B \subset\subset X$ can be left unmarried at stage α if and only if $|\{i \notin I_\beta : F(i) \subseteq Z\}| + |B| \leq m_\beta(Z)$ for every $\beta < \alpha$ and every $Z \supseteq B$ with $m_\beta(Z) \in \omega$, and this last condition means that $B \subset\subset X$ can be omitted at stage α. Hence $l_\alpha(X) = m_\alpha(X)$ by Theorem 8.6.

Now we present a variation of the criterion of [HPS2]. Remember that we have defined I_α by recursion as follows:

$I_0 = \emptyset$ and $I_\alpha = \cup\{I_\beta : \beta < \alpha\}$,

if α is a limit ordinal. If α is a successor ordinal $\beta + 1$, then

$I_\alpha = \{i : \text{There are an } A \subset\subset F(i) \text{ and a set } K \subseteq I_\beta \text{ critical in } F \setminus A$ such that $F(i) \subseteq F(K) \cup A\}$.

Let X be a set of women. Instead of counting the women of X who are left unmarried after we have married I_α, we now are counting those men in I_α who demand a woman of X if we marry I_α.

Since $I_0 = \emptyset$, no man of I_0 demands a woman of X if we marry $F \upharpoonright I_0$. Therefore let

$I_0(X) = \emptyset$.

Next we observe that $I_1 = \{i : F(i) \text{ is finite}\}$. If $F(i) \subset\subset X$, then clearly i demands a woman of X if we marry I_1. We put $I_1(X) = \{i : F(i) \subset\subset X\}$. In this case Hall's Theorem says that the following conditions are equivalent:

(1) $F \upharpoonright I_1$ has a marriage.
(2) $|I_1(B)| \leq |B|$ for every finite set B.

If α is an arbitrary ordinal, then the condition that $|\{i \in I_\alpha : F(i) \subseteq X\}| \leq |X|$ is necessary but not sufficient for the existence of a marriage of $F \upharpoonright I_\alpha$. In general there are more men in I_α who demand a woman of X. We proceed in the following way.

By the definition of I_β, $\beta\leq\alpha$, we can choose a Hall family $H = (H(i) : i\in I_\alpha)$ as follows.

If $i\in I_\alpha$, then there is a $\beta<\alpha$ with $i\in I_{\beta+1}\setminus I_\beta$, and we can choose $H(i)\subset\subset F(i)$ such that $F(i)\subseteq F(K)\cup H(i)$ for some $K\subseteq I_\beta$ critical in $F\setminus H(i)$. Moreover, if there are an $A\subset\subset F(i)\cap X$ and a set $K\subseteq I_\beta$ critical in $F\setminus A$ with $F(i)\subseteq F(K)\cup A$, then we can choose $H(i)\subset\subset X$. $H = (H(i) : i\in I_\alpha)$ is a skeleton of $F\restriction I_\alpha$. To see this, define $r : I_\alpha\to\alpha$ by $r(i) = \beta$ if $i\in I_{\beta+1}\setminus I_\beta$ as a rank function for H. We call H a <u>canonical skeleton of $F\restriction I_\alpha$</u>.

By Theorem 6.5, the family $F\restriction I_\alpha$ has a marriage if and only if $H\in\mathcal{M}$. Hence every $i\in I$ with $H(i)\subseteq X$ demands a woman of X if we marry $F\restriction I_\alpha$. So we define

$$I_\alpha(X) = \{i : \text{There are a } \beta<\alpha,\text{ an } A\subset\subset F(i)\cap X,\text{ and a set } K\subseteq I_\beta \text{ critical in } F\setminus A \text{ such that } i\notin I_\beta \text{ and } F(i)\subseteq F(K)\cup A\}.$$

Then we have:

1. $I_\alpha(F(\mathrm{dom}\,F)) = I_\alpha$.
2. $\{i\in I_\alpha : F(i)\subseteq X\}\subseteq I_\alpha(X)$.
3. If H is a canonical skeleton of $F\restriction I_\alpha$, then $\{i\in I_\alpha : H(i)\subseteq Y\}\subseteq I_\alpha(Y)$ for every set Y.
4. For any set X, there is a canonical skeleton H of $F\restriction I_\alpha$ such that $\{i\in I_\alpha : H(i)\subseteq X\} = I_\alpha(X)$.

From these facts we get:

<u>Theorem 8.8</u>.
The following conditions are equivalent:

(1) $F\restriction I_\alpha\setminus A$ has a marriage.
(2) $|I_\alpha(B)|\leq|B\setminus A|$ for every finite set B.

<u>Proof</u>. Assume (1), let B be a finite set, and let H be a canonical skeleton of $F\restriction I_\alpha$ such that $\{i : H(i)\subseteq B\} = I_\alpha(B)$. Since $(F\restriction I_\alpha)\setminus A\in\mathcal{M}$, $(H\restriction I_\alpha)\setminus A\in\mathcal{M}$ by Theorem 6.5, and therefore $|B\setminus A|\geq|H(I_\alpha(B))\setminus A|\geq|I_\alpha(B)|$.

Conversely let H be a canonical skeleton of $F \restriction I_\alpha$, and let $L \subset\subset I_\alpha$. By Hall's Theorem, it is sufficient to show that $|H(L) \setminus A| \geq |L|$. If we take $B = H(L)$, then $L \subseteq I_\alpha(B)$. Therefore

$$|H(L) \setminus A| \geq |B \setminus A| \geq |I_\alpha(B)| \geq |L|.$$

Now we can reformulate some results of Section 7.

Corollary 8.9.
The following conditions are equivalent:
(1) $F \restriction I_F$ has a marriage.
(2) $|I_\alpha(B)| \leq |B|$ for every finite set B and every ordinal α.

Proof. Since $I_\alpha \subseteq I_F$ by definition, (1) implies (2) by Theorem 8.8. Conversely we have $I_F = I_\gamma$ by Theorem 7.4. Therefore $F \restriction I_\gamma =$ $= F \restriction I_F$, and (2) implies (1) by Theorem 8.8 again.

Corollary 8.10.
If F is a countable family, then the following conditions are equivalent:
(1) F has a marriage.
(2) $|I_\alpha(B)| \leq |B|$ for every finite set B and every ordinal α.

Proof. Since F is countable, $\operatorname{dom} F \setminus I_F$ is countable. Hence the corollary follows from Corollary 7.6 and Corollary 8.9.

Corollary 8.11.
The following conditions are equivalent:

(1) F is critical.
(2) (i) $|I_\alpha(B)| \leq |B|$ for every ordinal α and every finite set B, and
(ii) For each $b \in F(\operatorname{dom} F)$, there are an ordinal α and a finite set B such that $b \in B$ and $|I_\alpha(B)| = |B|$.

Proof. By Theorem 7.4, we have $I_F = I_\gamma = I_\alpha$ for each $\alpha > \gamma$. Assume that F is critical. Then $F \in \mathcal{M}$ and $\operatorname{dom} F = I_F = I_\gamma$ by Lemma 7.3. Now (i) holds by Theorem 8.8. If $b \in F(\operatorname{dom} F)$, then $F \restriction I_\gamma \setminus \{b\}$ has no marriage. Therefore Theorem 8.8 implies that there is a set B with $|I_\gamma(B)| > |B \setminus \{b\}|$. Then $|I_\gamma(B)| = |B|$ by (i), and consequently $b \in B$.

Now assume that (2) is satisfied for F. By (i) and Theorem 8.8, $F \upharpoonright I_\gamma \in \mathcal{M}$. The Main Lemma implies that there is a set $L \subseteq I_\gamma$ critical in F such that $F \upharpoonright I_\gamma \setminus \{b\}$ has a marriage for each $b \notin F(L)$. If $b \in F(\operatorname{dom} F)$, then $F \upharpoonright I_\gamma \setminus \{b\}$ has no marriage by Theorem 8.8 and by (ii). So $F(\operatorname{dom} F) \subseteq F(L)$, and $\operatorname{dom} F \subseteq I_{\gamma+1} = I_\gamma$. Since $F \upharpoonright I_\gamma$ has a marriage, $L = \operatorname{dom} F$.

We want to show that we can define $I_\alpha(X)$ by recursion without using the notion of a critical set. For this reason we want to strengthen Theorem 8.8. We need the following technical lemma:

Lemma 8.12. Let $B \subset\subset Y$ and $K \subseteq I_\alpha$ critical in $F \setminus B$. Then there is a set $L \subseteq I_\alpha \setminus I_\alpha(Y)$ critical in $F \setminus Y$ such that $F(K) \subseteq F(L) \cup Y$.

Proof. We prove the lemma by induction on α.
Let L be maximal critical in $F \upharpoonright (I_\alpha \setminus I_\alpha(Y)) \setminus Y$. We want to show that $F(K) \cup B \subseteq F(L) \cup Y$. Since $B \subseteq Y \cup F(L)$, there is, by Fact 3, a set $N \subseteq K$ critical in $F \setminus (F(L) \cup Y)$ such that $F(N) \cup F(L) \cup Y = F(K) \cup F(L) \cup Y$. By Fact 2, $(N \cup L)$ is critical in $F \setminus Y$. If $N \subseteq I_\alpha \setminus I_\alpha(Y)$, then we get, by the maximality of L, $N \subseteq L$ and so $F(K) \subseteq F(L) \cup Y$. So it remains to show that $N \cap I_\alpha(Y) = \emptyset$. Let $i \in I_\alpha(Y)$. Then there are a $\beta < \alpha$, an $A \subset\subset Y$, and a set $M \subseteq I_\beta$ critical in $F \setminus A$ such that $F(i) \subseteq F(M) \cup A$. By the induction hypothesis, there is a set $M' \subseteq I_\beta \setminus I_\beta(Y) = I_\beta \setminus I_\alpha(Y) \subseteq I_\alpha \setminus I_\alpha(Y)$ critical in $F \setminus Y$ such that $F(i) \subseteq F(M) \cup A \subseteq F(M') \cup Y$. By the maximality of L, it follows that $F(i) \subseteq F(M') \cup Y \subseteq F(L) \cup Y$, and so $i \notin N$.

Corollary 8.13. Let $X \subseteq Y$, and let H be a canonical skeleton of $F \upharpoonright I_\alpha$. Then $(H \setminus I_\alpha(X)) \setminus Y$ is a skeleton of $(F \upharpoonright I_\alpha \setminus I_\alpha(X)) \setminus Y$.

Proof. Let $i \in I_\alpha \setminus I_\alpha(X)$. Then there are a $\beta < \alpha$ and a $K \subseteq I_\beta$ critical in $F \setminus H(i)$ with $i \in I_\beta$ and $F(i) \subseteq F(K) \cup H(i)$. By Lemma 8.12, there is a set $L \subseteq I_\beta \setminus I_\beta(Y \cup H(i)) \subseteq I_\beta \setminus I_\beta(X) \subseteq I_\alpha \setminus I_\alpha(X)$ critical in $F \setminus (Y \cup H(i)) = (F \setminus Y) \setminus (H(i) \setminus Y)$ such that $F(L) \cup Y \cup H(i) \supseteq F(K)$. So $(F \setminus Y)(i) \subseteq (F \setminus Y)(L) \cup (H(i) \setminus Y)$. This proves Corollary 8.13.

Theorem 8.14.

Let $X \subseteq Y$. Then the following conditions are equivalent:

(a) $(F \restriction I_\alpha \setminus I_\alpha(X)) \setminus Y$ has a marriage.

(b) $|I_\alpha(D) \setminus I_\alpha(X)| \leq |D \setminus Y|$ for every finite set D.

Proof. Assume (a) and let H be a canonical skeleton of $F \restriction I_\alpha$ with $I_\alpha(D) = \{i : H(i) \subseteq D\}$. By Corollary 8.13, $(H \setminus I_\alpha(X)) \setminus Y$ is a skeleton of $(F \restriction I_\alpha \setminus I_\alpha(X)) \setminus Y$. Thus, by Theorem 6.5 and (a), $(H \setminus I_\alpha(X)) \setminus Y \in \mathcal{M}$. So $|D \setminus Y| \geq |H(I_\alpha(D) \setminus I_\alpha(X)) \setminus Y| \geq |I_\alpha(D) \setminus I_\alpha(X)|$.

To prove the converse, let H be a canonical skeleton of $F \restriction I_\alpha$. It is sufficient to show that $(H \setminus I_\alpha(X)) \setminus Y$ fulfils Hall's condition. Let $L \subset\subset I_\alpha \setminus I_\alpha(X)$ and let $D = H(L)$. Then $L \subseteq I_\alpha(D)$ by property 3 (p. 153). Thus $|H(L) \setminus Y| = |D \setminus Y| \geq |I_\alpha(D) \setminus I_\alpha(X)| \geq |L|$.

Theorem 8.15.

For each finite set A the following conditions are equivalent:

(1) There is no set $K \subseteq I_\alpha$ critical in $F \setminus A$ such that $X \subseteq F(K) \cup A$.

(2) There are a set $Y \supseteq A$ and an $a \in X \setminus Y$ such that $|I_\alpha(D) \setminus I_\alpha(Y)| \leq |D \setminus (Y \cup \{a\})|$ for every finite set D.

Proof. Assume (1), let L be maximal critical in $F \restriction I_\alpha \setminus A$ and put $Y = F(L) \cup A$. By Fact 2, $\emptyset$ is the only critical set in $F \restriction I_\alpha \setminus Y$. Since $I_\alpha(Y) \supseteq \{i \in I_\alpha : F(i) \subseteq Y\}$, there is no K critical in $(F \restriction I_\alpha \setminus I_\alpha(Y)) \setminus Y$ and no $i \in I_\alpha \setminus (I_\alpha(Y) \cup K)$ such that $F(i) \setminus Y \subseteq F(K) \setminus Y$. By Corollary 8.13, $(F \restriction I_\alpha \setminus I_\alpha(Y)) \setminus Y$ has a skeleton, and so $(F \restriction I_\alpha \setminus I_\alpha(Y)) \setminus Y$ is stable. Thus $(F \restriction I_\alpha \setminus I_\alpha(Y)) \setminus Y \in \mathcal{M}$ by Corollary 7.9. Property (1) implies that there is an $a \in X \setminus Y$. Since $\emptyset$ is the only critical set in $(F \restriction I_\alpha \setminus I_\alpha(Y)) \setminus Y$, it follows, by the Main Lemma, that $(F \restriction I_\alpha \setminus I_\alpha(Y)) \setminus (Y \cup \{a\})$ has a marriage. Then, by Theorem 8.14, $|I_\alpha(D) \setminus I_\alpha(Y)| \leq |D \setminus (Y \cup \{a\})|$ for every finite set D.

Now let us assume property (2). By Theorem 8.14, $(F \restriction I_\alpha \setminus I_\alpha(Y)) \setminus (Y \cup \{a\}) \in \mathcal{M}$. Hence there is no $L \subseteq I_\alpha \setminus I_\alpha(Y)$ critical in $(F \restriction I_\alpha \setminus I_\alpha(Y)) \setminus Y$ with $a \in F(L) \cup Y$. Therefore, by Lemma 8.12, there is no $K \subseteq I_\alpha$ critical in $F \setminus A$ with $a \in F(K) \cup A$. So there is no K critical in $F \setminus A$ such that $X \subseteq F(K) \cup A$.

For each family F, we define by recursion on α demand functions J_α as follows: Put

$$J_\alpha(X) := \{i \in \mathrm{dom}(F) : i \text{ demands a woman of } X \text{ at stage } \beta \text{ for some } \beta < \alpha\}.$$

Then we say that <u>i demands a woman of X at stage α</u> if $i \notin J_\alpha(F(\mathrm{dom}\,F))$ and there is a set $B \subset\subset X \cap F(i)$ such that, for each $Y \supseteq B$ and every $a \in F(i)$, there is a finite set D such that $a \in D$ and $|J_\alpha(D) \setminus J_\alpha(Y)| \geq |D \setminus Y|$.

<u>Theorem 8.16</u>. $J_\alpha(X) = I_\alpha(X)$.

<u>Proof</u>. We prove the theorem by induction on α. By Theorem 8.15, $i \in I_\alpha(X)$ if and only if there are a $\beta < \alpha$ and a $B \subset\subset X \cap F(i)$ such that $i \notin I_\beta$ and, for each $a \in F(i)$ and each $Y \supseteq B$, there is a finite set D with $a \in D$ and $|I_\beta(D) \setminus I_\beta(Y)| \geq |D \setminus Y|$. By the induction hypothesis, we get $J_\beta(F(\mathrm{dom}\,F)) = I_\beta(F(\mathrm{dom}\,F)) = I_\beta$, $I_\beta(D) = J_\beta(D)$, and $I_\beta(Y) = J_\beta(Y)$. Hence $i \in I_\alpha(X)$ if and only if i demands a woman of X at stage β for some $\beta < \alpha$. This proves Theorem 8.16.

Chapter IV
MISCELLANEOUS THEOREMS ON MARRIAGES

§1. A special criterion

If $F = (F(i) : i \in I)$ is a family, then denote by $I_{<\aleph_0}$, I_∞ the sets $\{i \in I : |F(i)| < \aleph_0\}$, $\{i \in I : |F(i)| \geq \aleph_0\}$ respectively. In combination with Hall's Theorem it is natural to ask for conditions for the family $F \upharpoonright I_\infty$ which guarantee that the whole family F has a marriage. The following theorem of Milner and Shelah [MS2] gives an answer.

Theorem 1.1.
Let F be a family, and let κ be an infinite cardinal such that
(i) $|I_\infty| \leq \kappa$ and
(ii) $|F(i)| \leq \kappa$ for each $i \in I_\infty$.
Then F has a marriage if and only if every subfamily F^* of F with $|F^*| \leq \kappa$ has a marriage.

Proof. Of course every subfamily of F has a marriage if F has a marriage.

To prove the converse, let us define, for every set $N \subseteq F(I)$ and every finite subset B of N, a set $G_N(B)$ by

$$G_N(B) = \{J \subset\subset I_{<\aleph_0} : N \cap F(J) = B \wedge |F(J) \setminus B| < |J| \wedge \forall i \in J\ F(i) \nsubseteq N\}.$$

Now we define recursively a sequence $(A_n : n < \omega)$ of subsets of $F(I)$. Put $A_0 = F(I_\infty)$ and let A_n be defined. To each set $B \subset\subset A_n$ we construct a set $H_n(B) \subseteq F(I)$ as follows. If $G_{A_n}(B) \neq \emptyset$, then choose a set $J \subset G_{A_n}(B)$ and put $H_n(B) := F(J)$. If $G_{A_n}(B) = \emptyset$, then put $H_n(B) = B$.
Now define $A_{n+1} = \cup\{H_n(B) : B \subset\subset A_n\}$.
Finally put $A := \cup\{A_n : n < \omega\}$, $I_1 = \{i \in I : F(i) \subseteq A\}$ and $I_2 = I \setminus I_1$. Of course $I_2 \subseteq I_{<\aleph_0}$. By construction, $|A_n| \leq \kappa$ for each $n < \omega$, and so $|A| \leq \kappa$.
Now, by Hall's Theorem, the family $F \upharpoonright I_{<\aleph_0}$ has a marriage g, and

clearly $g[I_{<\aleph_0} \cap I_1] \subseteq A$. Therefore $|I_{<\aleph_0} \cap I_1| \leq \varkappa$, and so it follows by assumption that $F \upharpoonright I_1$ has a marriage f. To get a contradiction, let us assume that $F \setminus f$ has no marriage. By Hall's Theorem, there is a set $K \subset\subset I_2$ such that $|F(K) \setminus \mathrm{rng}\, f| < |K|$. If $B = F(K) \cap A$, then $B \subset\subset A$ and $|F(K) \setminus B| < |K|$. There exists a natural number n_0 such that $B \subset\subset A_{n_0}$. $(A_n : n < \omega)$ is an increasing chain, so $B \subset\subset A_n$ for each $n \geq n_0$. Since $K \cap I_1 = \emptyset$, the set K is an element of $G_{A_n}(B)$ for each $n \geq n_0$. By the definition of A_{n+1}, one can choose a set $J_n \in G_{A_n}(B)$ for each $n \geq n_0$. If $n_0 \leq k < s$, then it follows by the definition of the set $G_{A_n}(B)$ that $J_k \cap J_s = \emptyset$. Since

$$|F(J_n) \setminus B| < |J_n|$$

and since $f \upharpoonright J_n$ is a marriage of $F \upharpoonright J_n$, there is an $i_n \subset J_n$ such that $f(i_n) \in B$ for each natural number $n \geq n_0$. Therefore B must be infinite, contradicting the fact that B is a finite set.

§ 2. Shelah families and Nash-Williams families

A family $F = (F(i) : i \in I)$ is called a Shelah family if there is an infinite regular cardinal $\varkappa \leq |I|$ such that $|F(i)| < \varkappa$ for all $i \in I$. F is said to be a Nash-Williams family if $F^{-1}(x) = \{i \in I : x \in F(i)\}$ is countable for all $x \in F(I)$. In this section, we prove Shelah's necessary and sufficient criterion for Shelah families (see [S2]) which was a very important step in establishing a condition for arbitrary families. For example, K.P.Podewski's μ-Test Criterion is a generalization of Shelah's result. Shelah's condition, as stated below, shows for Shelah families that it is only necessary to check μ-tests of some special kind. In the second part of this section we prove Theorem III.3.8 for Nash-Williams families (see [N4]).

Definition.

(i) If $F = (F(i) : i \in I)$ is a family, $\mu > \omega$ is a regular cardinal with $\mu \leq |I|$, and $(A_\alpha : \alpha < \mu)$ is a continuous chain such that $|A_\alpha| < \mu$ and $A_\alpha \subseteq F(I)$ for all $\alpha < \mu$, then we call $(A_\alpha : \alpha < \mu)$ a special μ-test of F.

(ii) A special μ-test $(A_\alpha : \alpha < \mu)$ is called <u>positive</u> if there is a club C in μ such that, for all $\alpha \in C$, the following holds:

(1) If $F(i) \subseteq A_\alpha$, then there is a $\beta < \alpha$ with $F(i) \subseteq A_\beta$.

(2) $F_{A_\alpha}^{A_\beta} \in \mathcal{M}$ for all $\beta \geq \alpha$.

(3) $F^{A_\alpha} \in \mathcal{M}$.

<u>Theorem 2.1</u>. (Shelah)
Let $F = (F(i) : i \in I)$ be a Shelah family. Then $F \in \mathcal{M}$ if and only if F satisfies Hall's condition, i.e. $|J| \leq |F(J)|$ for all $J \subset\subset I$, and if, for each uncountable regular cardinal $\mu \leq |I|$, every special μ-test of F is positive.

<u>Proof</u>. By the μ-Test Criterion, Shelah's condition is necessary for the existence of a marriage. To prove that Shelah's condition is sufficient, let $F = (F(i) : i \in I)$ be a Shelah family satisfying Hall's condition and assume that, for each uncountable regular cardinal $\mu \leq |I|$, every special μ-test of F is positive. If $|I| \leq \aleph_0$, then Theorem 2.1 follows by Hall's theorem. If $|I| = \lambda$ is a regular uncountable cardinal, then we can assume without loss of generality that $\operatorname{dom} F = \lambda$ and $F(I) \subseteq \lambda$. We shall prove that

$$B_1 = \{\alpha < \lambda : \exists i \in I \ (F(i) \subseteq \alpha \wedge \sup F(i) = \alpha)\} \text{ and}$$
$$B_2 = \{\alpha < \lambda : \exists A \subseteq \lambda \ (\alpha \subseteq A \wedge |A| < \lambda \wedge F_\alpha^A \notin \mathcal{M})\}$$

are not stationary in λ. To get a contradiction let us assume that B_1 is stationary. Then the special λ-test $(A_\alpha : \alpha < \lambda)$, defined by $A_\alpha = \alpha$, is not positive. This contradicts the hypothesis that every special λ-test of F is positive. To get a contradiction again, assume that B_2 is stationary. As before we define a negative special λ-test $(A_\alpha : \alpha < \lambda)$. Put $A_0 = \emptyset$ and $A_\alpha = \bigcup\{A_\beta : \beta < \alpha\}$ for all limit ordinals $\alpha < \lambda$. If $\alpha = \beta + 1$ is a successor ordinal, then we distinguish two cases.

<u>Case 1</u>. $A_\beta = \beta \in B_2$.

In this case let $A_{\beta+1}$ be a set $A \subseteq \lambda$ such that $\beta \subseteq A$, $|A| < \lambda$, and $F_{A_\beta}^A \notin \mathcal{M}$. Of course $\beta \subseteq A_{\beta+1}$.

<u>Case 2</u>. Not Case 1.

Let $A_{\beta+1}$ be the least ordinal $\sigma > \beta$ such that $\rho < \sigma$ for all $\rho \in A_\beta$. Again $\beta \subseteq A_{\beta+1}$.

Since $\alpha \subseteq A_{\alpha+1}$ for each $\alpha<\lambda$, and each A_α is a subset of some $\beta<\lambda$, the set $C = \{\alpha<\lambda : A_\alpha = \alpha\}$ of fixed points is a club in λ, and therefore $C \cap B_2$ is stationary in λ. Obviously $C \cap B_2 \subseteq \{\alpha<\lambda : F_{A_\alpha}^{A_{\alpha+1}} \notin \mathcal{M}\}$, and so it follows that $\{\alpha<\lambda : F_{A_\alpha}^{A_{\alpha+1}} \notin \mathcal{M}\}$ is stationary in λ. Therefore $(A_\alpha : \alpha<\lambda)$ is a special λ-test which is not positive, contradicting the hypothesis that every special λ-test of F is positive. This proves that also B_2 is not stationary.

Let $C \subseteq \lambda$ be a club in λ such that $C \cap (B_1 \cup B_2) = \emptyset$, and let $\varphi : \lambda \to C \cup \{0\}$ be a well-ordering isomorphism. If $i \in I$, then there is a least ordinal ξ such that $F(i) \subseteq \varphi(\xi)$. Assume for contradiction that ξ is a limit ordinal. Since $\varphi(\xi) \notin B_1$, there is an ordinal $\sigma < \varphi(\xi)$ and an ordinal $\rho<\xi$ such that $F(i) \subseteq \sigma$ and $\sigma = \varphi(\rho)$. This contradicts the minimality of ξ. Therefore ξ is a successor ordinal. It follows that

$$I = \bigcup\{\operatorname{dom} F_{\varphi(\alpha)}^{\varphi(\alpha+1)} : \alpha<\lambda\}.$$

Since $C \cap B_2 = \emptyset$, there is a marriage h_α of $F_{\varphi(\alpha)}^{\varphi(\alpha+1)}$ for all $\alpha<\lambda$, and so $h = \bigcup\{h_\alpha : \alpha<\lambda\}$ is a marriage of F.

For the singular case, i.e. the case that $\lambda = |I|$ is singular, we prove the existence of a marriage of F by induction on singular cardinals. Let $J \subseteq I$ and $|J|<\lambda$. We distinguish two cases: If $|J|$ is regular, then $F \restriction J$ has a marriage as we proved above. But if $|J|$ is singular, then $F \restriction J$ has a marriage by the inductive hypothesis. By Theorem II.5.11, F has a marriage. This proves Theorem 2.1.

To avoid the lengthy proof of Theorem 4.4 which we used to get the Compactness Theorem 5.11, we give a direct proof for the singular case. If $\mu<\lambda = |I|$ is an uncountable regular cardinal and f is a matching in F with $|f|<\mu$, then we call f μ-admissible if $F^X \setminus f \in \mathcal{M}$ for each set X of cardinality less than μ. First we note that $\emptyset$ is μ-admissible. Second, if $|F(i)|<\varkappa$ for each $i \in I$ and $\mu>\varkappa$ is a regular cardinal less than λ, then we claim that, for any μ-admissible matching f in F and for any subset J of I with $|J|<\mu$, there is a μ-admissible matching g in F such that $f \subseteq g$ and $J \subseteq \operatorname{dom} g$. To prove this, put $B = F(J \cup \operatorname{dom} f)$. Then $|B|<\mu$. We first show that there exists a set $C \supseteq B$ with $|C|<\mu$ such that $F_C^D \in \mathcal{M}$ for each set $D \supseteq C$ with $|D|<\mu$. Suppose not. Let us define a μ-test $(A_\alpha : \alpha<\mu)$ by: $A_0 = \emptyset$, $A_1 = B$, $A_\alpha = \bigcup\{A_\beta : \beta<\alpha\}$ if $\alpha<\mu$ is a limit ordinal, and $A_{\beta+1}$ is a set $D \supseteq A_\beta$ with $|D|<\mu$ such that $F_{A_\beta}^D \notin \mathcal{M}$. The existence of this negative special μ-test $(A_\alpha : \alpha<\mu)$ contradicts the assumption of the theorem.

Since f is μ-admissible and $|C|<\mu$, the family $F^C\setminus f$ has a marriage h. Therefore $g = f\cup h$ is a marriage of F^C such that $J\subseteq \mathrm{dom}\, g$, $f\subseteq g$, and $|g|<\mu$.

If X is a set with $|X|<\mu$, then, by the construction of the set C, $F_C^{C\cup X}$ has a marriage g^*. Since g is a marriage of F^C, $\mathrm{rng}\, g\subseteq C$, and so the matching $g^*\restriction (I\setminus \mathrm{dom}\, g)$ is a marriage of $F^{C\cup X}\setminus g$. Therefore $F^X\setminus g\in \mathcal{M}$, and thus g is a μ-admissible matching in F.

The proof for the existence of a marriage of F is the same as the proof given on page 57. For convenience we repeat it. Let $\nu = cf(\lambda)$, and let $(\nu_\alpha : \alpha<\nu)$ be an increasing sequence of ordinals such that $\lambda = \sup\{\nu_\alpha : \alpha<\nu\}$ and $\max\{\kappa,\nu\}<\nu_\alpha$ for each $\alpha<\nu$. Let $(J_\alpha : \alpha<\nu)$ be a continuous chain such that $I = \bigcup\{J_\alpha : \alpha<\nu\}$ and $|J_\alpha|\leq \nu_\alpha$ for each $\alpha<\nu$. Since, for every μ-admissible matching f in F and every subset J of I with $|J|<\mu$, there exists a μ-admissible matching g in F such that $J\subseteq \mathrm{dom}\, g$ and $f\subseteq g$, it is possible to define, for each $\alpha<\nu$, a sequence $(g_n^\alpha : n<\omega)$ of matchings in F such that the following conditions are fulfilled:

(1) If $\alpha<\nu$, then $g_n^\alpha\subseteq g_{n+1}^\alpha$ for each $n<\omega$.

(2) If $\alpha<\nu$, then g_n^α is a ν_α^+-admissible matching in F such that $J_\alpha\subseteq \mathrm{dom}\, g_n^\alpha$.

(3) If $\alpha<\beta<\nu$ and $g_n^\beta(i)\in \mathrm{rng}\, g_n^\alpha$, then $i\in \mathrm{dom}\, g_{n+1}^\alpha$.

Now put $g^\alpha = \bigcup\{g_n^\alpha : n<\omega\}$ for each $\alpha<\nu$. Then

(i) $J_\alpha\subseteq \mathrm{dom}\, g^\alpha$.

(ii) If $\alpha<\beta<\nu$ and $g^\beta(i)\in \mathrm{rng}\, g^\alpha$, then $i\in \mathrm{dom}\, g^\alpha$.

If $L_\alpha = \mathrm{dom}\, g^\alpha\setminus \bigcup\{\mathrm{dom}\, g^\gamma : \gamma<\alpha\}$ for any $\alpha<\nu$, then, by properties (i) and (ii),

$$h = \bigcup\{g^\alpha\restriction L_\alpha : \alpha<\nu\}$$

is a marriage of F.

If $F = (F(i) : i\in I)$ with $\aleph_2\leq |I|$ is a Shelah family, i.e. if there is an infinite regular cardinal $\kappa\leq |I|$ such that $|F(i)|<\kappa$ for each $i\in I$, then one may conjecture that for the existence of a marriage of F it is enough to require that F is μ-positive for each uncountable regular cardinal $\mu<|I|$. The next example (see [MS2]), being a subtle modification of the example on page 45, shows that this conjecture is false. Put

$J_1 = \{\alpha<\aleph_2 : cf(\alpha) = \aleph_1\}$, $\quad J_2 = \{\beta<\aleph_1 : \mathrm{Lim}(\beta)\}$, and

$I = \{\alpha + \beta : \alpha \in J_1 \wedge \beta \in J_2\}$.

For any $\alpha \in J_1$, there is an increasing sequence of limit ordinals $s(\alpha,\beta)$ $(\beta < \aleph_1)$ such that $\alpha = \sup\{s(\alpha,\beta) : \beta < \aleph_1\}$. Define

$$F(\alpha + \beta) = ((\alpha + \beta) \setminus \alpha) \cup \{s(\alpha,\beta)\}$$

for each ordinal $\alpha + \beta \in I$. Since the family which we discussed on page 45 has no marriage, this family has no marriage, too. Obviously $|F(i)| < \aleph_1$ for all $i \in I$. We will show that every subfamily F' of F with $|F'| \leq \aleph_1$ has a marriage. It is enough to prove that each subfamily F_ρ of F has a marriage, where

$$F_\rho = (F(\alpha + \beta) : \alpha \in J_1 \wedge \beta \in J_2 \wedge \alpha < \rho\})$$

and $\rho < \aleph_2$. Let ρ be fixed. If $|\{\alpha \in J_1 : \alpha < \rho\}| = \nu$, then let $(\alpha_\tau : \tau < \nu)$ be an enumeration of $\{\alpha \in J_1 : \alpha < \rho\}$ such that $\alpha_\sigma \neq \alpha_\tau$ whenever $\sigma < \tau < \aleph_1$. Put

$$A_\tau = \{s(\alpha_\tau, \beta) : \beta < \aleph_1\}$$

for each $\tau < \nu$. Obviously $|A_\tau \cap A_\sigma| \leq \aleph_0$ for any $\tau < \sigma < \nu$, and so each set $B_\tau = A_\tau \setminus \bigcup\{B_\sigma : \sigma < \tau\}$, $\tau < \nu$, is non-empty. For each $\tau < \nu$ put $\beta_\tau = \min\{\beta : s(\alpha_\tau, \beta) \in B_\tau\}$ and let f be a marriage of the family $(\beta : \beta \in J_2 \wedge \beta < \beta_\tau)$. The function g with $\operatorname{dom} g = \operatorname{dom} F_\rho$, defined by

$$g(\alpha_\tau + \beta) = \begin{cases} \alpha_\tau + f_\tau(\beta) + 1 & \text{if } \beta < \beta_\tau, \\ s(\alpha_\tau, \beta) & \text{if } \beta_\tau \leq \beta \end{cases},$$

is a marriage of F_ρ.

In Theorem 2.1, we cannot drop the requirement that F is a Shelah family:

Let $F : \omega_1 \cup \{-1\} \to \omega_1 \cup \{\omega_1\}$ be defined by $F(-1) = \omega_1$ and $F(\alpha) = \{\alpha\}$ for each $\alpha < \omega_1$. This family satisfies Shelah's condition but has no marriage.

Now we want to eliminate the conditions "$F_{A_\alpha}^{A_\beta} \in \mathcal{M}$" and "$F^{A_\alpha} \in \mathcal{M}$" in the definition of a special μ-test. For this reason we introduce the notion of a κ-intact family. Let κ be an uncountable cardinal. A family $F = (F(i) : i \in I)$ is said to be <u>κ-intact</u>, if either

(1) $|F| < \kappa$ and $F \in \mathcal{M}$

or

(2) $|F| \geq \varkappa$, F satisfies the extended Hall condition, i.e. $|J| \leq |F(J)|$ for each $J \subseteq I$, and, for each uncountable regular cardinal μ with $\mu \leq |I|$ and each sequence $(A_\alpha : \alpha < \mu)$ with $A_\alpha \subseteq F(I)$ and $|A_\alpha| < \mu$ for each $\alpha < \mu$, there exists a club C in μ such that for all $\alpha \in C$ the following holds:

(i) If $F(i) \subseteq A_\alpha$, then there is a $\beta < \alpha$ with $F(i) \subseteq A_\beta$.

(ii) $F_{A_\alpha}^{A_\beta}$ is $\varkappa$-intact for all $\beta \geq \alpha$.

(iii) F^{A_α} is $\varkappa$-intact.

<u>Theorem 2.2</u>. Let $\varkappa$ be an uncountable regular cardinal, and let $F = (F(i) : i \in I)$ be a family such that $|F(i)| < \varkappa$ for any $i \in I$. Then $F \in \mathcal{M}$ if and only if F is $\varkappa$-intact.

<u>Proof</u>. The proof is by transfinite induction on $|I|$. If $|I| < \varkappa$, then Theorem 2.2 is obviously true. Therefore let $F = (F(i) : i \in I)$ be a family such that $|F(i)| < \varkappa \leq |I|$ for all $i \in I$ (i.e. F is a Shelah family), and let $F \in \mathcal{M}$. Then F satisfies the extended Hall condition. To prove that F is $\varkappa$-intact, let $\mu > \omega$ be a regular cardinal with $\mu \leq |I|$ and let $(A_\alpha : \alpha < \mu)$ be a sequence with $A_\alpha \subseteq F(I)$ and $|A_\alpha| < \mu$ for each $\alpha < \mu$. By Theorem 2.1, there exists a club C in μ such that, for each $\alpha \in C$, the following holds:

(i) If $F(i) \subseteq A_\alpha$, then there is a $\beta < \alpha$ with $F(i) \subseteq A_\beta$.

(ii) $F_{A_\alpha}^{A_\beta} \in \mathcal{M}$ for all $\beta \geq \alpha$.

(iii) $F^{A_\alpha} \in \mathcal{M}$.

Let $\alpha \in C$. F satisfies the extended Hall condition, so $|F^{A_\alpha}| \leq |A_\alpha| < \mu \leq |I|$. Since $|F_{A_\alpha}^{A_\beta}(i)| \leq |F^{A_\beta}(i)| \leq |F(i)| < \varkappa$ for $\alpha < \mu$ and $\beta \geq \alpha$, it follows by the inductive hypothesis that $F_{A_\alpha}^{A_\beta}$ and F^{A_α} are $\varkappa$-intact. This proves that F is $\varkappa$-intact.

To prove the converse, suppose that F is $\varkappa$-intact. We shall verify that F fulfils the condition of Theorem 2.1. So let $\mu > \omega$ with $\mu \leq |I|$ be a regular cardinal and $(A_\alpha : \alpha < \mu)$ be a sequence with $A_\alpha \subseteq F(I)$ and $|A_\alpha| < \mu$ for each $\alpha < \mu$. Since F is $\varkappa$-intact, there is a club C in μ such that, for each $\alpha \in C$, the following holds:

(i) If $F(i) \subseteq A_\alpha$, then there is a $\beta < \alpha$ such that $F(i) \subseteq A_\beta$.

(ii) $F_{A_\alpha}^{A_\beta}$ is $\varkappa$-intact for each $\beta \geq \alpha$.

(iii) F^{A_α} is $\varkappa$-intact.

Since F satisfies the extended Hall condition, it follows that $|F_{A_\alpha}^{A_\beta}| < \mu \leq |I|$ and $|F^{A_\alpha}| < \mu \leq |I|$. By the inductive hypothesis, $F_{A_\alpha}^{A_\beta} \in \mathcal{M}$ and $F^{A_\alpha} \in \mathcal{M}$. Therefore every μ-test $(A_\alpha : \alpha < \mu)$ of F is positive, and so $F \in \mathcal{M}$ by Theorem 2.1.

Now we give the proof that any Nash-Williams family $F = (F(i) : i \in I)$ has a marriage if and only if $F(i) \not\subseteq F(K)$ for each set K critical in F and each $i \in I \setminus K$. The proof requires the following simple lemma.

<u>Lemma 2.3</u>. Every Nash-Williams family with uncountable members only has a marriage.

<u>Proof</u>. Let $F^* = (F^*(i) : i \in I)$ be a Nash-Williams family such that $|F^*(i)| > \aleph_0$ for all $i \in I$. Choose, for each $i \in I$, a subset $F(i)$ of $F^*(i)$ such that $|F(i)| = \aleph_1$. Every component of the bipartite graph Γ_F associated with $F = (F(i) : i \in I)$ has cardinality $\aleph_1$: For any $i_0 \in I$ put $J_0 = \{i_0\}$ and $J_{n+1} = \{i \in I : F(i) \cap F(J_n) \neq \emptyset\}$. Then $|J_n| \leq \aleph_1$, and so $|\bigcup\{J_n : n \in \omega\}| \leq \aleph_1$.

Now let $\Gamma_{F \upharpoonright J}$ be a connected component of Γ_F and $(j_\alpha : \alpha < \varkappa \leq \aleph_1)$ be an enumeration of J. Choose, for every $\alpha < \varkappa$, $x_\alpha \in F(j_\alpha) \setminus \{x_\beta : \beta < \alpha\}$. Then the map h, defined by $h(j_\alpha) = x_\alpha$ for each $\alpha < \varkappa$, is a marriage of $F \upharpoonright J$. This proves $F \in \mathcal{M}$.

<u>Theorem 2.4</u>. (Nash-Williams [N4])
A Nash-Williams family $F = (F(i) : i \in I)$ has a marriage if and only if $F(i) \not\subseteq F(K)$ for all sets K critical in F and all $i \in I \setminus K$ (i.e. if F is not 1-obstructed).

<u>Proof</u>. Of course the condition of Theorem 2.4 is necessary for the existence of a marriage. To prove the converse, let $F = (F(i) : i \in I)$ be a Nash-Williams family such that $F(i) \not\subseteq F(L)$ for each set $L \subseteq I$ which is critical in F and each $i \in I \setminus L$. By Theorem

II.2.4, there is a set $K \subseteq I$ which is maximal critical in F. The family F^*, defined by

$$F^* = (F \setminus K) \setminus F(K),$$

has no non-empty critical subfamily. Obviously F^* is a Nash-Williams family. It is enough to prove that $F^* \in \mathcal{M}$. Put

$I_0 = \{i \in I \setminus K : |F^*(i)| > \aleph_0\}$ and

$I_1 = I \setminus (K \cup I_0)$.

Every connected component of the bipartite graph $\Gamma_{F^* \restriction I_1}$ associated with $F^* \restriction I_1$ is countable. Therefore there is a partition $(J_\alpha : \alpha < \varkappa)$ of I_1 with the following properties:

(1) $|J_\alpha| \le \aleph_0$ for each $\alpha < \varkappa$.

(2) $|F^*(J_\alpha)| \le \aleph_0$ for each $\alpha < \varkappa$.

(3) $F^*(J_\alpha) \cap F^*(J_\beta) = \emptyset$ whenever $\alpha < \beta < \varkappa$.

Let us define a new family $\tilde{F} = (\tilde{F}(i) : i \in I_0)$ by

$$\tilde{F}(i) = \{F^*(J_\alpha) : F(i) \cap F^*(J_\alpha) \neq \emptyset\} \cup (F^*(i) \setminus F^*(I_1)).$$

By property (2), it follows that $|\tilde{F}(i)| > \aleph_0$ for all $i \in I_0$. Again by property (2), $\tilde{F}$ is a Nash-Williams family. Lemma 2.3 yields a marriage h of $\tilde{F}$. If $h(i) = F^*(J_\alpha)$ for some $\alpha < \varkappa$, then choose an element $x_i \in F(i) \cap F^*(J_\alpha)$. Otherwise put $x_i = h(i)$. The function g, defined by $g(i) = x_i$ for each $i \in I_0$, is a marriage of $F^* \restriction I_0$. Note that for each $\alpha < \varkappa$ there is at most one $x_i \in F^*(J_\alpha)$.

$F^* \restriction J_\alpha$ is a countable family which has no non-empty critical subfamily. By the Corollaries III.3.5 and III.3.9, there is a marriage f_α of $F^* \restriction J_\alpha$ for each $\alpha < \varkappa$. If $x_i \in F^*(J_\alpha)$, then, by Corollary II.2.12, it is possible to choose a marriage f_α of $F^* \restriction J_\alpha$ such that $x_i \notin \operatorname{rng} f_\alpha$. This proves that $g \cup \bigcup\{f_\alpha : \alpha < \varkappa\}$ is a marriage of F^*.

§3. A sufficient criterion of Milner and Shelah

Let F be a family and $x \in F(I)$. Remember that $F^{-1}(x)$ denotes the set $\{j \in I : x \in F(j)\}$.

Definition. A family $F = (F(i) : i \in I)$ of nonempty sets is called a Milner-Shelah family if $|F^{-1}(x)| \le |F(i)|$ whenever $i \in I$ and $x \in F(i)$.

If Γ_F is the bipartite graph associated with F, then $d(i) = |F(i)|$ and $d(x) = |F^{-1}(x)|$ denote the degree of i and x respectively. A family F of non-empty sets is a Milner-Shelah family if $d(x) \leq d(i)$ whenever $i \in I$ and $x \in F(i)$. In other words, if a man i knows a woman x, then the number of men whom the woman x knows is not greater than the number of women whom the man i knows. It turns out that every Milner-Shelah family has a marriage. But the following simple example shows that not every family which has a marriage is a Milner-Shelah family. Define F by $I = \{0,1\}$, $F(0) = \{0,1\}$, $F(1) = \{1\}$. The woman 1 knows the men 0 and 1, but the man 1 only knows the woman 1; so F is not a Milner-Shelah family, but $f = \{(0,0),(1,1)\}$ is a marriage of F.

Lemma 3.1. If F is a finite, critical Milner-Shelah family and $x \in F(i)$, then $d(i) = d(x)$.

Proof. We prove the lemma by induction on the number of edges of the bipartite graph Γ_F associated with F. If $E(\Gamma_F) = \emptyset$, then Lemma 3.1 is obviously true. So let $E(\Gamma_F) \neq \emptyset$.

Case 1. $F(i) = \{a\}$ for some $i \in I$ and some $a \in F(I)$.

If $H = F\setminus\{i\}\setminus\{a\}$, then H is critical. Since F is a Milner-Shelah family, it follows that

(1) $d_F(i) = 1 = d_F(a)$.

(2) $d_H(j) = d_F(j)$ for each $j \in I \setminus \{i\}$ and
$d_H(b) = d_F(b)$ for each $b \in F(I) \setminus \{a\}$.

Thus H is a critical Milner-Shelah family, and therefore Lemma 3.1 follows from the induction hypothesis, (1), and (2).

Case 2. $F(i) \neq \{a\}$ for all $i \in I$, $a \in F(I)$.

Let f be a marriage of F and $H = (F(i) \setminus \{f(i)\} : i \in I)$. Since $f(I) = F(I)$, we get

(3) $d_F(j) = d_H(j) + 1$ for each $j \in I$ and
$d_F(b) = d_H(b) + 1$ for each $b \in F(I)$.

So H is a Milner-Shelah family. We will show that H is critical, and then Lemma 3.1 follows from the induction hypothesis and (3). It is sufficient to prove that H has a marriage. If $H \notin \mathcal{M}$, then there are a set K critical in H and an $i \in I \setminus K$ such that $H(i) \subseteq H(K)$. Take $a \in H(i)$ and $k \in K$ such that $a \in H(i) \cap H(K)$. Then $d_H(a) > d_{H \restriction K}(a)$. Since H is a Milner-Shelah family, $H \restriction K$ is a Milner-Shelah family, too. Thus, by the inductive hypothesis, $d_{H \restriction K}(a) = d_{H \restriction K}(k) = d_H(k)$, and therefore $d_H(a) > d_H(k)$, which contradicts the fact that H is a Milner-Shelah family.

Corollary 3.2. If F is a finite, critical Milner-Shelah family such that the bipartite graph Γ_F associated with F is connected, then Γ_F is regular, i.e. there is a natural number n such that $d(i) = n = d(x)$ for all $i \in I$ and $x \in F(I)$.

Corollary 3.3. Every Milner-Shelah family F satisfies Hall's condition, i.e. $|J| \leq |F(J)|$ for all $J \subset\subset F(I)$.

Proof. To get a contradiction, let us assume that there is a finite set $J \subseteq I$ such $|F(J)| < |J|$. Let J_0 be a $\subseteq$-minimal set with this property. Choose an element $j_0 \in J_0$. Then $|F(J_0 \setminus \{j_0\})| = |J_0 \setminus \{j_0\}|$ by the minimality of J_0. Therefore, by Hall's Theorem, $F \upharpoonright J_0 \setminus \{j_0\}$ is a critical family and $F(j_0) \subseteq F(J_0 \setminus \{j_0\})$. Since F is a Milner-Shelah family, $F(j_0) \neq \emptyset$. Therefore $J_0 \setminus \{j_0\} \neq \emptyset$. Choose an element $k \in J_0 \setminus \{j_0\}$ such that $F(k) \cap F(j_0) \neq \emptyset$. If $H = F \upharpoonright J_0 \setminus \{j_0\}$ and $a \in F(k) \cap F(j_0)$, then

$$d_H(k) = d_H(a)$$

by Lemma 3.1, and so

$$d_F(k) < d_F(a).$$

This contradicts the hypothesis that F is a Milner-Shelah family.

Definition. A matching f in a family F with $|f| < \aleph_0$ is called admissible if the family $F \setminus f$ satisfies Hall's condition.

Lemma 3.4. If F is a Milner-Shelah family, f an admissible matching in F, and $J \subset\subset I$ a finite subset of the domain of F, then there is an admissible matching g in F such that $g \supseteq f$ and $J \subseteq \mathrm{dom}\, g$.

Proof. It is enough to prove Lemma 3.4 for singletons $J = \{i_0\}$. Let f be admissible. If $i_0 \in \mathrm{dom}\, f$, then put $g = f$. If $i_0 \notin \mathrm{dom}\, f$, then define $I_0 = \{i \in I : |F(i)| < \aleph_0\}$. Since $F \setminus f$ satisfies Hall's condition, so does the family $(F \upharpoonright I_0) \setminus f$. If $i_0 \in I_0$, then let h be a marriage of $(F \upharpoonright I_0) \setminus f$. The function $g := f \cup \{(i_0, h(i_0))\}$ is a matching in F such that $F \setminus g$ satisfies Hall's condition. Therefore $g \supseteq f$ is an admissible matching such that $i_0 \in \mathrm{dom}\, g$.

Now we discuss the case that $i_0 \notin I_0$. The family $H = F \upharpoonright (I_0 \cup \{i_0\} \cup \mathrm{dom}\, f)$ is a Milner-Shelah family which has only finitely many infinite members. Therefore

(1) $d_H(x) < \aleph_0$ for each $x \in H(\mathrm{dom}\, H)$.

It is enough to show that $H \setminus f$ has a marriage h. Then $g = f \cup \{(i_0, h(i_0))\}$ is an admissible matching in F with the desired properties. - By (1), $t = \sum_{x \in \mathrm{rng}\, f} d_H(x)$ is a finite cardinal.

Since $|F(i_0)| \geq \aleph_0$, there exists a finite set $B \subseteq F(i_0) \setminus \operatorname{rng} f$ with $|B| > t$. Put $H^* = (H \restriction I_0) \setminus f = (F \restriction I_0) \setminus f$. We claim that there is a marriage h^* of the family H^* such that $B \not\subseteq \operatorname{rng} h^*$. Clearly $H^* \in \mathcal{M}$. To get a contradiction, let us assume that $B \subseteq \operatorname{rng} h^*$ for each marriage h^* of H^*. By Theorem III.1.6, there is a finite set $K \subseteq \operatorname{dom} H^*$ critical in H^* such that $B \subseteq H^*(K)$. If

$n = \sum_{k \in K} |H^*(k)| = \sum_{x \in H^*(K)} d_{H^* \restriction K}(x)$, then $\sum_{k \in K} |F(k) \cap \operatorname{rng} f| =$

$= (\sum_{k \in K} |F(k)|) - n$. By the definition of t, we get $t \geq (\sum_{k \in K} |F(k)|) - n$. If g^* is a marriage of $H^* \restriction K$, then $\sum_{k \in K} |F(k)| \geq \sum_{k \in K} d_F(g^*(k)) \geq$

$\geq |B| + \sum_{k \in K} d_{F \restriction K}(g^*(k)) \geq |B| + \sum_{k \in K} d_{H^* \restriction K}(g^*(k)) \geq |B| + \sum_{x \in H^*(K)} d_{H^* \restriction K}(x) =$

$= |B| + n$. It follows that $(\sum_{k \in K} |F(k)| - n) \geq |B|$, and so $t \geq |B| > t$. This contradiction shows that there is a marriage h^* of H^* such that $B \not\subseteq \operatorname{rng} h^*$. Choose an element $b \in B \setminus \operatorname{rng} h^*$ and put $h = h^* \cup \{(i_0, b)\}$. h is a marriage of $H \setminus f$.

Lemma 3.5. If F is a Milner-Shelah family such that $|I_\infty| \leq \aleph_0$, then $F \in \mathcal{M}$. Especially every countable Milner-Shelah family has a marriage.

Proof. Let $F = (F(i) : i \in I)$ be a Milner-Shelah family, $I_{<\aleph_0} = \{i \in I : |F(i)| < \aleph_0\}$, $I_\infty = \{i \in I : |F(i)| \geq \aleph_0\}$, and $(i_k : k < n \leq \omega)$ an enumeration of the set I_∞. We define recursively a sequence $(f_k : k < n \leq \omega)$ with the following properties:

1. f_k is a marriage of $F \restriction \{i_l : l \leq k\}$.
2. $f_r \subseteq f_s$, if $r < s < n$.
3. f_k is an admissible matching.

Put $f_0 = \emptyset$. By Corollary 3.3, f_0 is an admissible matching of F. Let f_l with properties 1. - 3. be defined for each $l < k < n$. Since f_{k-1} is an admissible matching, there exists, by Lemma 3.4, an admissible matching $f_k \supseteq f_{k-1}$ of $F \restriction \{i_l : l \leq k\}$.

The function $f = \bigcup\{f_k : k < n \leq \omega\}$ is a marriage of $F \restriction I_\infty$. To prove that $F \setminus f$ satisfies Hall's condition, let J be a finite subset of $I_{<\aleph_0}$. Then $(F \setminus f)(J) = F(J) \setminus \operatorname{rng} f_k$ for some $k < n$. By property 3, $|J| \leq |F(J) \setminus \operatorname{rng} f_k| = |(F \setminus f)(J)|$.

Therefore $F\setminus f$ satisfies Hall's condition, and so there is a marriage g of $(F\setminus I_{\infty})\setminus \operatorname{rng} f$. The function $h = f\cup g$ is a marriage of F.

Theorem 3.6. Every Milner-Shelah family has a marriage.

Proof. Let $F = (F(i): i\in I)$ be a Shelah-Milner family. We prove Theorem 3.6 by induction on $|I|$. If $|I|\leq\aleph_0$, then Theorem 3.6 is proved by Lemma 3.5. For $|I|>\aleph_0$ we distinguish two cases.

Case 1. $|I|>\aleph_0$ is regular.

Put $\lambda := |I|$ and $B := \{x\in F(I): d(x)<\lambda\}$.

Furthermore define

$I_{<\lambda} = \{i\in I: |F(i)|<\lambda\}$ and

$I_{\geq\lambda} = \{i\in I: |F(i)|\geq\lambda\}$.

For each $i\in I_{\geq\lambda}$, choose a subset $F^*(i)\subseteq F(i)$ with $|F^*(i)| = \lambda$ such that the following properties hold:

1. If $|F(i)\cap B|\geq\lambda$, then $F^*(i)\subseteq F(i)\cap B$.
2. If $|F(i)\cap B|<\lambda$, then $F^*(i)\subseteq F(i)\setminus B$.

Define

$F_1 = (F^*(i): i\in I_{\geq\lambda} \wedge |F(i)\cap B|<\lambda)$,

$F_2 = (F(i): i\in I_{<\lambda})\cup(F^*(i): i\in I_{\geq\lambda} \wedge |F(i)\cap B|\geq\lambda)$

Since F is a Milner-Shelah family, it follows by the definition of the set B that $F(I_{<\lambda})\subseteq B$. Therefore $F_2(\operatorname{dom} F_2)\subseteq B$. Since $F_1(\operatorname{dom} F_1)\subseteq F(I)\setminus B$, we have only to prove that F_1 and F_2 have a marriage. But since $|F^*(i)| = \lambda$ for every $i\in\operatorname{dom} F_1$, the family F_1 has a marriage. Therefore it remains to prove that the Milner-Shelah family F_2 has a marriage.

To simplify the notation let us assume that $F = (F(i): i\in I)$ is a Milner-Shelah family with $|I|\leq\lambda$ such that $d(x)<\lambda$ for all $x\in F(I)$. On the set $\bar{\mathcal{M}}$ of all Milner-Shelah families $(G(i): i\in I)$ such that $G(i)\subseteq F(i)$ for each $i\in I$, we define a relation $\leq$ by the following rule: $(G_1(i): i\in I)\leq(G_2(i): i\in I)$ if and only if $G_2(i)\subseteq G_2(i)$ and $(G_1(i)\subsetneq G_2(i) \Rightarrow |G_1(i)|<|G_2(i)|)$ for all $i\in I$. If $(G_\alpha:\alpha<\kappa)$ is a decreasing chain of elements of $\bar{\mathcal{M}}$, then

$G := (\cap\{G_\alpha(i):\alpha<\kappa\}: i\in I)$

is a Milner-Shelah family, $G\in\bar{\mathcal{M}}$, and G is a lower bound of $(G_\alpha:\alpha<\kappa)$. By Zorn's Lemma, $\bar{\mathcal{M}}$ has a minimal element. So let us assume without loss of generality that F is minimal with respect to $\leq$.

<u>Claim</u>. $|F(i)|<\lambda$ for all $i\in I$.

Suppose that the claim is proved. For any $i\in I$ put $J_0 = \{i\}$, $J_{n+1} = \{i\in I : \exists x\in F(J_n)\ x\in F(i)\}$, and $J = \cup\{J_n : n<\omega\}$. Since λ is regular, it follows by the claim that $|J|<\lambda$. By the inductive hypothesis, $F\restriction J$ has a marriage. Therefore every component of Γ_F has a marriage, and so the family F has a marriage.

We prove the claim. To get a contradiction, let us assume that there is an element $i\in I$ such that $|F(i)|\geq\lambda$. Choose a well-ordering $<$ on $F(i)$ with the following property: If $x,y\in F(i)$ and $x<y$, then $d(x)\leq d(y)$. For $x\in F(i)$ denote by

$$\hat{x} = \{y\in F(i) : y<x\}$$

the set of predecessors of x. First of all we prove that $|\hat{x}|<d(x)$ for all $x\in F(i)$. To get a contradiction, let us assume that there is an element $x\in F(i)$ such that $d(x)\leq|\hat{x}|$. Choose a subset $Y\subseteq\hat{x}$ with $|Y| = d(x)$ and replace the member $F(i)$ of F by the set Y. If $y\in Y$, then $d(y)\leq d(x)<|F(i)|$ and therefore this new family is a Milner-Shelah family, contradicting the minimality of the family F. Now we define by recursion a sequence $(x_n : n<\omega)$ of elements of $F(i)$ with the following properties:

1. $x_0<x_1<x_2<\ldots<x_n<\ldots$.
2. $d(x_n) = |\hat{x}_{n+1}|$.

Let x_0 be the $<$-smallest element of $F(i)$ and let $x_0,\ldots,x_n$ be defined. Since $|\hat{x}_n|<d(x_n)<\lambda$, there is an element $x_{n+1}\in F(i)$ such that $|\hat{x}_{n+1}| = d(x_n)$. It follows that $x_n<x_{n+1}$.

Since λ is regular, the cardinal $\rho = \sup|\hat{x}_n| = \sup d(x_n)$ is less than λ. If $x = \sup x_n$, then $|\hat{x}| = \rho$. We want to prove that there is an element $z<x$ such that $d(z)>\rho$. To get a contradiction, let us assume that $d(z)\leq\rho$ for each $z<x$. Now replace the member $F(i)$ of the family F by $\hat{x} = \{z\in F(i) : z<x\}$. The resulting family is a Milner-Shelah family, and this contradicts the minimality of F. Therefore there is an element $z<x$ such that $d(z)>\rho$. Choose a natural number n such that

$$z<x_n<x.$$

Then

$$\rho<d(z)\leq d(x_n)\leq\rho.$$

This contradiction proves that $|F(i)|<\lambda$ for each $i\in I$.

Case 2. $|I|$ is singular.

Put $|I| = \lambda$ and $\varkappa = cf(\lambda)$. Since λ is a singular cardinal, there is a sequence $(J_\alpha : \alpha < \varkappa)$ of pairwise disjoint subsets of I such that $I = \cup\{J_\alpha : \alpha < \varkappa\}$ and $|J_\alpha| < \lambda$ for each $\alpha < \varkappa$. Put $F_\alpha = F \restriction J_\alpha$ and partition J_α into two classes

$J_\alpha^0 = \{i \in J_\alpha : |F(i)| \leq \varkappa\}$ and

$J_\alpha^1 = \{i \in J_\alpha : |F(i)| > \varkappa\}$.

For each $i \in J_\alpha^1$, let $(F_\alpha(i,\beta) : \beta < \varkappa)$ be a family of pairwise disjoint sets such that $F_\alpha(i) = \cup\{F_\alpha(i,\beta) : \beta < \varkappa\}$ and $|F_\alpha(i,\beta)| = |F_\alpha(i)|$ for all $\beta < \varkappa$. The family

$$\Gamma_\alpha^* = F_\alpha \restriction J_\alpha^0 \cup (F_\alpha(i,\beta) : i \in J_\alpha^1 \wedge \beta < \varkappa)$$

is a Milner-Shelah family, and $|\mathrm{dom}\, F_\alpha^*| = |J_\alpha^0| + |J_\alpha^1| \cdot \varkappa < \lambda + \lambda \cdot \varkappa = \lambda$. By the inductive hypothesis, F_α^* has a marriage f_α. We now define a family $G = (G(i) : i \in I)$ by

$$G(i) = \begin{cases} F(i) & \text{if } |F(i)| \leq \varkappa, \\ \{f_{\alpha(i)}(i,\beta) : \beta < \varkappa\} & \text{if } |F(i)| > \varkappa, \end{cases}$$

where $\alpha(i)$ denotes the ordinal $\alpha < \varkappa$ such that $i \in J_\alpha$. Note that $G(i) \subseteq F(i)$ for all $i \in I$.

Claim. G is a Milner-Shelah family.

If $x \in G(I)$, then, for each $\alpha < \varkappa$, there is at most one $i \in J_\alpha$ and one $\beta < \varkappa$ such that $x = f_{\alpha(i)}(i,\beta)$. Therefore $d_G(x) \leq d_F(x)$. Now let $x \in G(i)$ and $i \in J_\alpha$. If $i \in J_\alpha^0$, then $G(i) = F(i)$ and $d_G(x) \leq d_F(x) \leq |F(i)| = |G(i)|$. Now let $i \in J_\alpha^1$. If there is an ordinal $\delta < \varkappa$ and an element $j \in J_\delta^0$ such that $x \in F(j)$, then $d_G(x) \leq d_F(x) \leq |F(j)| \leq \varkappa = |G(i)|$. Otherwise $d_G(x) \leq \varkappa = |G(i)|$. This proves the claim.

Since $|G(i)| \leq \varkappa$ for each $i \in I$, every component of the bipartite graph Γ_G associated with G has a cardinality less than or equal to $\varkappa$. By the induction hypothesis, every component of Γ_G has a marriage, and so it follows that G has a marriage. Since $G(i) \subseteq F(i)$ for each $i \in I$, the family F has a marriage.

§4. Cotransversals

If $F = (F(i) : i \in I)$ is a family and f a marriage of F, then we call $u(f) = F(I) \setminus \operatorname{rng} f$ a <u>cotransversal</u> of F. The question whether the supremum of the set

$$\{|u(f)| : f \text{ is a marriage of } F\}$$

is a maximum, was first solved by M.Ziegler. We here present, with the author's kind permission, an unpublished proof by H.A.Jung.

<u>Theorem 4.1</u>. Let F be a family which has a marriage. Then the set

$$\{|u(f)| : f \text{ is a marriage of } F\}$$

is a closed interval of cardinal numbers.

<u>Proof</u>. Let f, g be marriages of F and $\varkappa$ be a cardinal such that $|u(f)| < \varkappa < |u(g)|$. We are looking for a marriage h of F such that $|u(h)| = \varkappa$.

Every component of the bipartite graph

$$\Gamma = (I, F(I), f \cup g)$$

is a finite path, a two-way infinite path, a one-way infinite path or a cycle of even length. Let p be a component of Γ which is a finite path. If $p = (v)$, then $v \in u(f) \cap u(g)$, and if $p = (v_0, v_1)$ and $v_0 \in F(I)$, then $f(v_1) = v_0 = g(v_1)$. Finally if $p = (v_0, v_1, \ldots, v_n)$ with $n \geq 2$, $v_0 \in F(I)$, and $f(v_1) = v_0$, then $v_0 \in u(g)$ and $v_n \in u(f)$. So $|V(p) \cap u(f)| = |V(p) \cap u(g)|$.

If $\varkappa$ is finite, then there are at least $\varkappa - |u(f)|$ one-way infinite paths p_α which are a component of Γ and contain exactly one element of $u(g)$. Then, if $f \Delta E(p_\alpha) = (f \setminus E(p_\alpha)) \cup (E(p_\alpha) \setminus f)$,

$$h = \bigcup\{f \Delta E(p_\alpha) : \alpha < \varkappa - |u(f)|\}$$

is a marriage of F with $|u(h)| = \varkappa$.

But if $\varkappa$ is infinite, then there are $\varkappa$ one-way infinite paths p_α which are a component of Γ and contain exactly one element of $u(g)$. Then $h = \bigcup\{f \Delta E(p_\alpha) : \alpha < \varkappa\}$ is a marriage of F with $|u(h)| = \varkappa$. This proves that $\{|u(f)| : f \text{ is a marriage of } F\}$ is an interval.

Now let $\tilde{\mathcal{M}}$ be a set of marriages of F such that $\sup\{|u(f)| : f \in \tilde{\mathcal{M}}\} = \varkappa$. Suppose for contradiction that $|u(f)| < \varkappa$ for all marriages f of F. Choose $f_0 \in \tilde{\mathcal{M}}$.

Case 1. $\varkappa = \omega$.

We define by recursion a sequence $(f_n : n<\omega)$ of marriages of F and a sequence $(p_n : 1\leq n<\omega)$ such that p_n is an f_n-alternating path starting with an edge of f_n such that $f_{n+1} = f_n \Delta E(p_{n+1})$.

Let the marriages $f_0, f_1, \ldots, f_n$ of the family F and the paths $p_1, \ldots, p_n$ be defined. Choose a marriage $g \in \tilde{\mathcal{M}}$ such that $|u(g)| > n^2 + |u(f_n)|$.

As noticed above, there are in the bipartite graph $\Gamma = (I, F(I), g \cup f_n)$ n^2+1 components of Γ which are one-way infinite paths containing exactly one element of $u(g)$. At least one of these paths, call it p^*, does not meet the paths p_i $(1\leq i\leq n)$ in the first n vertices. Put

$$p_{n+1} = p^* \text{ and } f_{n+1} = f_n \Delta E(p^*).$$

Now let $i \in I$. If there is no natural number n such that $i \in V(p_n)$, then put $f(i) = f_0(i)$. Otherwise there is, by construction, a maximal natural number m such $i \in V(p_{m+1})$; put $f(i) = f_{m+1}(i)$. The function f is a marriage of F and every end vertex of the path p_n is an element of $u(f)$. Therefore $|u(f)| = \omega$.

Case 2. $\varkappa > \omega$.

We define by transfinite recursion a sequence $(p_\alpha : \alpha < \varkappa)$ of f_0-alternating, vertex disjoint, one-way infinite paths which start with an edge of f_0. Let p_β for $\beta<\alpha<\varkappa$ be defined. Since $|\alpha|\cdot\omega < \varkappa$, there exists a marriage $g \in \tilde{\mathcal{M}}$ such that $|u(g)| > \max\{|\alpha+1|\cdot\omega, |u(f_0)|\}$. In

$$\Gamma = (I, F(I), f_0 \cup g)$$

there are $|u(g)|$ components of Γ which are one-way infinite paths containing exactly one element of $u(g)$. Thus there exists one of these components, called p^*, such that $V(p^*) \cap \bigcup\{V(p_\beta) : \beta<\alpha\} = \emptyset$. Put $p_\alpha = p^*$. If $f = f_0 \Delta \bigcup\{E(p_\alpha) : \alpha < \varkappa\}$, then f is a marriage of F such that every end vertex of p_α is an element of $u(f)$. Therefore $|u(f)| = \varkappa$.

Bibliography

[A1] Aharoni,R.: On the equivalence of two conditions for the existence of transversals. J.Combin.Theory, Ser.A, 34 (1983), 202 - 214.

[A2] Aharoni,R.: Menger's Theorem for Graphs Containing no Infinite Paths. Europ.J.Combinatorics 4 (1983), 201 - 204.

[A3] Aharoni,R.: König's duality theorem for infinite bipartite graphs. J.London Math.Soc. 29 (1984), 1 - 12.

[A4] Aharoni,R.: Matchings in graphs of size $\aleph_1$. J.Combin. Theory, Ser.B, 36 (1984), 113 - 117.

[ANS1] Aharoni,R., Nash-Williams,C.St.J.A., and Shelah,S.: A general criterion for the existence of transversals. Proc.London Math.Soc. 47 (1983), 43 - 68.

[ANS2] Aharoni,R., Nash-Williams,C.St.J.A., and Shelah,S.: Another form of a criterion for the existence of transversals. J.London Math.Soc. 29 (1984), 193 - 203.

[BM] Bollobas,B. and Milner,E.C.: A theorem in transversal theory. Bull.London Math.Soc. 5 (1973), 267 - 270.

[BS] Brualdi,R.A. and Scrimger,E.B.: Exchange systems, matchings and transversals. J.Combin.Theory, Ser.B, 5 (1968), 244 - 257.

[DM] Damerell,M.R. and Milner,E.C.: Necessary and sufficient conditions for transversals of countable set systems. J.Combin.Theory, Ser.A, 17 (1974), 350 - 374.

[EGR] Erdös,P., Galvin,F., and Rado,R.: Transversals and multi-transversals. J.London Math.Soc. 20 (1979), 387 - 395.

[F] Fodor,G.: Eine Bemerkung zur Theorie der regressiven Funktionen. Acta Sci. Math. (Szeged) 17 (1956), 139 - 142.

[H.,P.] Hall,P.: On representatives of subsets. J.London Math.Soc. 10 (1935), 26 - 30.

[H.,M.] Hall,M.: Distinct representatives of subsets. Bull.Amer. Math.Soc. 54 (1948), 922 - 926.

[HPS1] Holz,M., Podewski,K.-P., and Steffens,K.: Hall families and the marriage problem. J.Combin.Theory, Ser.A, 27 (1979), 161 - 180.

[HPS2] Holz,M., Podewski,K.-P., and Steffens,K.: A Hall criterion for countable families of sets. J.London.Math.Soc. 21 (1980), 1 - 12.

[M1] Milner,E.C.: Transversal theory. Proceedings of the International Congress of Mathematicians (Vancouver 1974), Vol.1, 155 - 169.

[M2] Milner,E.C.: Transversals of infinite set systems. Second International Conference on Combinatorial Mathematics (New York 1978), Ann. New York Acad. Sci. 319, 397 - 401.

[M3] Milner,E.C.: Lectures on the marriage theorem of Aharoni, Nash-Williams and Shelah. Graph Theory, Singapore 1983. Proceedings (eds. K.M.Koh and H.P.Yap), Lecture Notes in Mathematics Nr.1073, 55 - 79.

[MS1] Milner,E.C. and Shelah,S.: Sufficiency conditions for the existence of transversals. Canad.J.Math. 26 (1974), 948 - 961.

[MS2] Milner,E.C. and Shelah,S.: Some theorems on transversals. Infinite and finite sets. Vol.III, Coloqu.Math.Soc.Janos Bolyai, Vol.10, North Holland, Amsterdam (1975), 1115-1126.

[Mi] Mirsky,L.: Transversal theory. Acad.Press, New York, London (1971).

[N1] Nash-Williams,C.St.J.A.: Which infinite set-systems have transversals? A possible approach. Proceedings of the conference on combinatorics at Oxford. Institute of mathematics and it's application (1972), 237-253.

[N2] Nash-Williams,C.St.J.A.: Marriage in denumerable societies. J.Combin.Theory, Ser.A, 19 (1975), 335-366.

[N3] Nash-Williams,C.St.J.A.: Another criterion for marriage in denumerable societies, Advances in graph theory (ed. B.Bollabas). Ann. Discrete Math. 3 (1978), 165-179.

[N4] Nash-Williams,C.St.J.A.: Marriage in societies in which each woman knows countably many men. Proc. of the tenth southeastern conference on combinatorics, graph theory and computing, Vol.1 (eds. F.Hoffman, D.McCarthy, R.C.Mullin, R.G.Stanton), Congressus Numerantium XXIII, Utilitas Mathematica, Publishing Inc. (1979), 103-115.

[OS] Oellrich,H. and Steffens,K.: On Dilworth decomposition theorem. Discrete Math. 15 (1976), 301-304.

[P] Podewski,K.-P.: Necessary and sufficient conditions for the existence of injective choice functions. Schriftenreihe des Instituts für Mathematik Nr.165, Universität Hannover (1983).

[PS1] Podewski,K.-P. and Steffens,K.: Injective choice functions for countable families, J.Combin.Theory. Ser.B, 21 (1976), 40-46.

[PS2] Podewski,K.-P. and Steffens,K.: Maximal representable subfamilies. Bull.London Math.Soc. 8 (1976), 186-189.

[PS3] Podewski,K.-P. and Steffens,K.: Über Translationen und den Satz von Menger in unendlichen Graphen. Acta Math.Acad.Sci. Hungar. 30 (1977), 69-84.

[S1] Shelah,S.: A substitute for Hall's theorem for families with infinite sets. J.Combin.Theory, Ser.A, 16 (1974), 199-208.

[S2] Shelah,S.: Notes on partition calculus, in infinite and finite sets, Vol.III, Colloq.Math.Soc.Janos Bolyai, North Holland, Amsterdam (1975), 1257-1276.

[S3] Shelah,S.: A compactness theorem for singular cardinals, free algebras, Whitehead problem and transversals. Isr.J. Math. 21 (1975), 319-349.

[St1] Steffens,K.: Injective choice functions. J.Combin.Theory, Ser.A, 17 (1974), 138-144.

[St2] Steffens,K.: Ergebnisse aus der Transversalentheorie I,II. J.Combin.Theory, Ser.A, 20 (1976), 178-229.

[St3] Steffens,K.: Faktoren in unendlichen Graphen. Jber.d.Dt. Math.-Verein. 87 (1985), 127-137.

[T] Tverberg,H.: On the Milner-Shelah condition for transversals. J.London Math. 13 (1976), 520-524.

[vdW] Van der Waerden,B.L.: Ein Satz über Klasseneinteilungen von endlichen Mengen. Abh.math.Sem.Hamburg.Univ. 5 (1927), 185-188.

INDEX OF SYMBOLS

GENERAL INDEX